# SPECTRAL THEORY AND COMPLEX ANALYSIS

NORTH-HOLLAND
**MATHEMATICS STUDIES** **4**

**Notas de Matemática (49)**
Editor: Leopoldo Nachbin
*Universidade Federal do Rio de Janeiro*
*and University of Rochester*

# Spectral Theory and Complex Analysis

**JEAN PIERRE FERRIER**
*University of Nancy I*

1973

NORTH-HOLLAND PUBLISHING COMPANY - AMSTERDAM • LONDON
AMERICAN ELSEVIER PUBLISHING COMPANY, INC. - NEW YORK

Library of Congress Catalog Card Number: 72 93089
ISBN North-Holland:
Series: 0 7204 2700 2
Volume: 0 7204 2704 5
ISBN American Elsevier: 0 444 10429 1

PUBLISHERS:

NORTH-HOLLAND PUBLISHING COMPANY - AMSTERDAM
NORTH-HOLLAND PUBLISHING COMPANY, LTD. - LONDON

SOLE DISTRIBUTORS FOR THE U.S.A. AND CANADA:

AMERICAN ELSEVIER PUBLISHING COMPANY, INC.
52 VANDERBILT AVENUE
NEW YORK, N.Y. 10017

PRINTED IN THE NETHERLANDS

# INTRODUCTION

These notes are issued from lectures given by the author at the "Collège de France" in 1971, the purpose of which was an exposition of complex analysis in $\mathbf{C}^n$ based on spectral theory. Such an approach leads to global theorems in connection with holomorphic convexity, approximation problems or ideals of holomorphic functions, and makes possible the introduction of growth conditions.

It is easy to apply the holomorphic functional calculus of Banach algebras to polynomial approximation of holomorphic functions on a neighbourhood of a polynomially convex compact set $K$ in $\mathbf{C}^n$, by proving that $K$ is the joint spectrum of the coordinates in the closed subalgebra generated by the polynomials in $\mathcal{E}(K)$. This method leads to the so-called Oka-Weil theorem. We remark that polynomial convexity is equivalent to the existence of a family $(p_\alpha)$ of polynomials such that

$$(1) \qquad 1/\chi_K(s) = \sup |p_\alpha(s)|$$

for every $s$ in $\mathbf{C}^n$, where $\chi_K$ denotes the characteristic function of $K$. As the holomorphic functional calculus only requires convexity with respect to the restrictions of polynomials to a given neighbourhood of $K$, an improvement consists in asking for condition (1) when $s$ belongs to such a neighbourhood.

It would be more difficult, however, to use the theory of Banach algebras in proving the same result when polynomials are replaced by the algebra $\mathcal{O}(\Omega)$ of holomorphic functions on a given pseudoconvex domain $\Omega$, and polynomial convexity by convexity with respect to $\mathcal{O}(\Omega)$, or the family $(p_\alpha)$ by a family $(f_\alpha)$ of $\mathcal{O}(\Omega)$. Moreover, the result is true upon replacing $|f_\alpha|$ by a positive function $\pi_\alpha$ on $\Omega$ such that $\log \pi_\alpha$ is plurisubharmonic, as it has been proved by L. Hörmander.

All theorems mentioned above concern approximation on $K$ of functions defined on a neighbourhood of $K$, or approximation for the compact open topology. It is possible to obtain better results by considering growth conditions, but the algebras of holomorphic functions will no longer be Banach algebras. We shall therefore use spectral theory of b-algebras, as introduced by L. Waelbroeck. A sufficiently general setting, including all classical examples, is the following : if $\delta$ is a non negative function on $\mathbf{C}^n$ such that $|\delta(s) - \delta(s')| \leqslant |s - s'|$ for all $s$, $s'$ in $\mathbf{C}^n$ and $|s|\,\delta(s)$ is uniformly bounded, we consider the algebra $\mathcal{C}(\delta)$ of complex functions $f$ on the open set $\{\delta > 0\}$ such that $\delta^N |f|$ is uniformly bounded for some positive integer $N$, and the subalgebra $\mathcal{O}(\delta)$ of all functions of $\mathcal{C}(\delta)$ which are holomorphic. Elementary properties of such algebras are given in Chapter I. We say that two non negative functions $\delta_1$, $\delta_2$ on $\mathbf{C}^n$

are equivalent if $\varepsilon\delta_1^N \leqslant \delta_2$ and $\varepsilon\delta_2^N \leqslant \delta_1$ for some positive integer N and some $\varepsilon>0$; if $\delta_1$, $\delta_2$ are equivalent, the algebras $\mathcal{O}(\delta_1)$, $\mathcal{O}(\delta_2)$ are equal.

Let $\delta$, $\delta'$ be two non negative functions on $\mathbf{C}^n$ satisfying the required properties. If $\delta' \geqslant \delta$ and if $\{\delta' > 0\}$ is connected, the algebra $\mathcal{O}(\delta')$ can be considered as a subalgebra of $\mathcal{O}(\delta)$. We say that $\mathcal{O}(\delta')$ is dense in $\mathcal{O}(\delta)$ if, for every function f of $\mathcal{O}(\delta)$, there exists a sequence $f_n$ in $\mathcal{O}(\delta')$ and a positive integer N such that $\delta^N |f_n - f|$ tends uniformly to zero. The basic approximation theorem, proved in Chapter VI, states that if $\{\delta' > 0\}$ is assumed to be pseudoconvex and, up to equivalence, $-\log\delta$ plurisubharmonic on $\{\delta>0\}$, a necessary and sufficient condition for density of $\mathcal{O}(\delta')$ in $\mathcal{O}(\delta)$ is the existence of a family $(f_\alpha)$ of $\mathcal{O}(\delta')$ satisfying, up to equivalence, the relation

$$1/\delta = \sup_\alpha |f_\alpha| \qquad (2)$$

on $\{\delta' > 0\}$ (resp. $\{\delta>0\}$). Moreover, the result is still valid upon replacing $f_\alpha$ by a positive function $\pi_\alpha$ of $\mathcal{C}(\delta')$ such that $\log \pi_\alpha$ is plurisubharmonic.

We apply our theorem to approximate holomorphic functions on a given domain by polynomials or holomorphic functions defined on a larger domain or satisfying more restrictive growth conditions. We discuss the Runge property through this method in Chapter VI .

The approximation theorem is not trivial when $\delta \neq \delta'$. When $\delta$ is the distance to the boundary of an open set, an easy consquence is the fact that every pseudoconvex domain is a domain of holomorphy; moreover, there exists an holomorphic function with polynomial growth which cannot be extended to a larger domain. Actually, these properties are deduced in Chapter IV from a theorem of I. Cnop concerning the joint spectrum of the coordinates in the algebra $\mathcal{O}(\delta)$. This spectral theorem appears as a particular case, but with parameter, of the corona problem for algebras of holomorphic functions with restricted growth, which has been solved by L. Hörmander.

Conversely, spectral theory has applications to the theory of ideals in algebras of holomorphic functions, which are given in Chapter V. Instead of $\mathcal{O}(\delta)$, we consider, more generally, algebras of holomorphic functions on a given domain $\Omega$ which are inductive limits of algebras $\mathcal{O}(\delta)$. We prove for such an algebra A the following decomposition property : if f vanishes at a point s of $\Omega$ , we have

$$f = (z_1 - s_1)g_1 + \ldots + (z_n - s_n)g_n ,$$

where $g_1, \ldots, g_n$ can be chosen in a bounded set of A when f varies in a bounded set of A and s in $\Omega$ . We discuss holomorphic convexity for algebras satisfying the decomposition property and give a characterization of inductive limits of algebras $\mathcal{O}(\delta)$.

Our exposition heavily depends on the estimates of L. Hörmander for the $\bar\partial$ Neumann problem, but only through the spectral theorem. All other properties are deduced by means of spectral theory and functional calculus. In Chapter VII , however, such estimates are used again to obtain new spectral and approximation theorems. We apply these last results to study plurisubharmonic functions on a pseudoconvex domain and give a generalization of a theorem of H. Bremermann which is independent of methods of Hartogs.

The study of algebras of bounded holomorphic functions on open domains is not suitable for spectral methods. In a quite different direction, it seems very difficult to introduce growth conditions in complex manifolds and develop $L^2$ - methods there. The case of relatively compact domains of complex manifolds have been studied by R. Narashiman, but it is not very different from the $\mathbf{C}^n$ case, because of the existence of a finite number of local maps.

The original exposition started from spectral theory of b - algebras and the unsimplified holomorphic functional calculus of L. Waelbroeck. It is not likely to assume that the reader knows everything about such a construction. Therefore, as an introduction to the theory, all definitions and properties which are essential for the following chapters are given in Chapter III, with a short construction of the functional calculus. The reader is referred to more complete texts for the multiplicative property. We discuss, however, the easier case of Banach algebras.

Algebras of holomorphic functions with restricted growth are not Banach algebras but directed unions of Banach spaces. It would have been possible to consider on such algebras the direct limit locally convex topology but bounded structures are more natural and lead to more precise theorems. Basic definitions about such structures are given in Chapter II.

A few elementary properties of plurisubharmonic functions and pseudoconvex domains are also recalled in Chapter IV. The basic estimates for the $\bar{\partial}$ Neumann problem are however admitted. The reader can find more information on these topics in the first chapters of Hörmander's book.

The original lectures were given in French; as I am not well - acquainted with the English language, the translation I have written on Publisher's request is probably awkward. However, I hope these notes will be useful to those who are looking for an introduction to a new aspect of complex analysis which is suitable for a large development.

Jean - Pierre Ferrier

Collège de France, Paris
January 1971

and

University of Nancy
October 1972

CONTENTS

# LIST OF SYMBOLS

$\chi_A$ is the characteristic function of A.
ker f is the kernel of the linear mapping f.
$\complement A$ is the complement of A.
$\bar{A}$, $\mathring{A}$, $\partial A$ are respectively the closure, the interior, the boundary of A.
z is the identity mapping of $\mathbf{C}^n$, $z_j$ the coordinate of index j.
$\{f > 0\}$, where f is a real function on $\mathbf{C}^n$ denotes the set of all $s \in \mathbf{C}^n$ such that $f(s) > 0$.
$|z|$ is the Euclidian norm in $\mathbf{C}^n$.
$d(s, A)$ is the Euclidian distance from a point s to a set A.
$d\lambda$ denotes the Lebesgue measure
$d''$ is the differential form $\partial/\partial\bar{z}_1 \, d\bar{z}_1 + \ldots + \partial/\partial\bar{z}_n \, d\bar{z}_n$.

$\mathcal{O}(\Omega)$ is the algebra of holomorphic functions in $\Omega$.
$\mathcal{C}(\delta)$, $\mathcal{O}(\delta)$ are defined in Section 1.1.
$\mathcal{C}(\Delta)$, $\mathcal{O}(\Delta)$ are defined in Section 1.5.
$\mathcal{C}(\delta; E)$, $\mathcal{C}_r(\delta; E)$, $\mathcal{CC}_r(\delta; E)$, $\mathcal{O}(\delta; E)$ are defined in Section 2.1.
${}_N\mathcal{C}(\delta)$, ${}_N\mathcal{C}_r(\delta)$, ${}_N\mathcal{CC}_r(\delta)$, ${}_N\mathcal{O}(\delta)$ are defined in Section 2.1.

$\delta_o = (1 + |z|^2)^{-1/2}$
$\delta_\Omega$ is defined in Section 1.2.
$\delta_{\Omega,\varphi}$, $\Delta_\Omega$, $\Delta_K$, $\Lambda_\varphi$, $\Lambda_{\Omega,\varphi}$ are defined in Section 1.5.
$\tilde{\delta}(s) = \inf_{s \in \mathbf{C}^n} (\delta(s') + |s' - s|)$ (see Section 1.4).

$E_B$ is the vector space spanned by B equipped with the Minkowski functional of B.
$\bar{F}$, when F is a vector subspace of a b-space E, is defined in Section 2.4.
$\hat{F}$ is defined in Section 2.3.

$A[X_1, \ldots, X_n]$ is defined in Section 2.5.
$\mathrm{idl}(a_1, \ldots, a_n; A)$ is the b-ideal generated by $a_1, \ldots, a_n$ in the b-algebra A (see Section 2.5).

$\mathrm{sp}(a) = \mathrm{sp}(a; A)$ (resp. $\mathrm{sp}(a_1, \ldots, a_n) = \mathrm{sp}(a_1, \ldots, a_n; A)$ is the spectrum of a (resp. the joint spectrum of $a_1, \ldots, a_n$) in A (see Section 3.1).

$\sigma(a) = \sigma(a;A)$ (resp. $\sigma(a_1,\dots,a_n) = \sigma(a_1,\dots,a_n;A)$) is the set of all spectral sets for a (resp. $a_1,\dots,a_n$) in A (see Section 3.2).

$\Delta(a_1,\dots,a_n) = \Delta(a_1,\dots,a_n;A)$ is the set of all spectral functions for $a_1,\dots,a_n$ in A (see Section 3.3).

$\Delta(a;A/I)$ is defined in Section 3.5.

$f[a]$ is defined in Section 3.4 (this is the holomorphic functional calculus at a).

$\mathbf{d}z = dz_1 \wedge \dots \wedge dz_n$, $\mathbf{d}''u = d''u_1 \wedge \dots \wedge d''u_n$.

$\langle x, y \rangle$ replaces $x_1y_1 + \dots + x_ny_n$

$\widehat{K}_\Omega$ is the hull of K with respect to plurisubharmonic functions in $\Omega$ (see Section 4.1).

$\widehat{\delta}$ is defined in Section 4.4.

${}^s\mathfrak{J}$ is defined in Section 5.1.

$\delta_f$ is defined in Section 5.3.

$\delta_B$ is defined in Section 5.4.

$\widehat{K}_H$, $\widehat{K}_P$, $\widehat{K}_\Omega$ are defined in Section 6.1.

$\widehat{K}_P$ is defined in Section 7.3.

$\nu(B)$ is the filtration of B (see Section 7.1).

$\widehat{\Omega}_P$, $\widehat{\delta}_P$ are defined in Section 7.5.

# CHAPTER I

## ALGEBRAS OF HOLOMORPHIC FUNCTIONS WITH RESTRICTED GROWTH

We define the algebra $\mathcal{T}(\delta)$ of tempered functions with respect to a weight function $\delta$, and the subalgebra $\mathcal{O}(\delta)$ of holomorphic functions of $\mathcal{T}(\delta)$. Weight functions are non negative functions on $\mathbf{C}^n$ such that $|s|\delta(s)$ is uniformly bounded on $\mathbf{C}^n$, satisfying $|\delta(s)-\delta(s')| \leqslant |s-s'|$ for all $s, s'$ in $\mathbf{C}^n$. Examples of such algebras are given, including the algebra of polynomials, entire functions of exponential type, entire functions of finite order or holomorphic functions with polynomial growth on an open set. A more general condition on weight functions is introduced, which actually leads to the same algebras; moreover, it can be assumed that each weight function $\delta$ is $C^\infty$ on the set $\{\delta>0\}$. We study inductive limits of algebras $\mathcal{O}(\delta)$ and the algebra of all holomorphic functions on a domain $\Omega$ .

### 1.1.- Basic definitions

Let $\delta$ be a non negative function on $\mathbf{C}^n$; we associate to $\delta$ the set $\{\delta>0\}$ where $\delta$ does not vanish and define $\delta$-tempered functions as complex valued functions on $\{\delta>0\}$ such that $\delta^N|f|$ is uniformly bounded on $\{\delta>0\}$ for some positive integer N. Thus $\delta$-tempered functions are functions on $\{\delta>0\}$ which are bounded by a positive multiple of some negative power of $\delta$ . The set of all $\delta$-tempered functionsis an algebra, which will be denoted by $\mathcal{T}(\delta)$.

We shall say that two non negative functions $\delta_1$, $\delta_2$ on $\mathbf{C}^n$ are equivalent if there exist a positive integer N and $\varepsilon>0$ such that

$$\varepsilon\delta_1^N \leqslant \delta_2 \quad \text{and} \quad \varepsilon\delta_2^N \leqslant \delta_1 .$$

Equivalent non negative functions give obviously rise to the same set and the same algebra of tempered functions. More generally, if $\varepsilon\delta_1^N \leqslant \delta_2$ for some positive integer

N and some $\varepsilon>0$, the set $\{\delta_2>0\}$ contains the set $\{\delta_1>0\}$, and the restriction mapping is an homomorphism from $\mathcal{T}(\delta_2)$ to $\mathcal{T}(\delta_1)$.

We suppose now that $\delta$ is a non negative function on $\mathbf{C}^n$ such that $\{\delta>0\}$ is open. For every positive integer r, we define $\mathcal{T}_r(\delta)$ as the algebra of all complex-valued functions on $\{\delta>0\}$ such that every derivative of order $s\leqslant r$ of f is $\delta$-tempered. We also define $\mathcal{O}(\delta)$ as the subalgebra of $\mathcal{T}(\delta)$ which consists of all functions in $\mathcal{T}(\delta)$ which are holomorphic on $\{\delta>0\}$. In other words

$$\mathcal{O}(\delta) = \mathcal{T}(\delta) \cap \mathcal{O}(\{\delta>0\}),$$

where $\mathcal{O}(\Omega)$ denotes the algebra of all holomorphic functions in the open set $\Omega$ ; we recall that holomorphic functions on $\Omega$ are exactly locally integrable functions on $\Omega$ satisfying $d''f = 0$ in the sense of distributions, where $d''$ is the differential operator

$$\partial/\partial \bar{z}_1 +\ldots+ \partial/\partial \bar{z}_n .$$

1.2.- Weight functions

In order to prove nice properties for $\mathcal{O}(\delta)$, we introduce restrictive conditions on $\delta$. Precisely, if $z = (z_1,\ldots,z_n)$ denotes the identity mapping of $\mathbf{C}^n$ and $|s| = (|s_1|^2+\ldots+|s_n|^2)^{\frac{1}{2}}$ the hermitian norm of s

Definition 1.- A non negative function $\delta$ on $\mathbf{C}^n$ is called a weight function if $\delta$ satisfies the following conditions :

W 1) $|z|\delta$ is uniformly bounded on $\mathbf{C}^n$.

W 2) $|\delta(s) - \delta(s')| \leqslant |s - s'|$ for all s, s' in $\mathbf{C}^n$.

Condition W 2 implies that $\delta$ is continuous. For a continuous $\delta$, condition W 1 means that $\delta = O(1/|z|)$ at infinity. Hence $\mathcal{O}(\delta)$ contains $1, z_1,\ldots, z_n$ and therefore all polynomials.

Condition W 2 is deeply connected with spectral theory, as we shall see in Chapter III. On condition W 2 also depend all the properties proved in the following section.

We consider now a few examples of weight functions which lead to classical algebras of holomorphic functions

1) Let

$$\delta_o = (1+|z|^2)^{\frac{1}{2}}.$$

The algebra $\mathcal{T}(\delta_o)$ consists of all complex valued functions defined on $\mathbf{C}^n$ which have polynomial growth at infinity. Thus $\mathcal{O}(\delta_o)$ is the algebra of polynomials, because of the theorem of Liouville.

We note that, for a continuous $\delta$, condition W 1 means that $\delta$ is bounded by a positive multiple of $\delta_o$.

2) If $\delta = e^{-|z|}$, the algebra $\mathcal{O}(\delta)$ consists of all entire functions f on $\mathbf{C}^n$ satisfying

$$e^{-N|z|} f = O(1),$$

for some positive integer N, that is

$$f = O(e^{N|z|}).$$

Thus $\mathcal{O}(e^{-|z|})$ is the algebra of entire functions of exponential type.

3) More generally, if k is a positive integer and $\varepsilon > 0$ so that $\varepsilon e^{-|z|^k}$ satisfies W 2, the algebra $\mathcal{O}(\varepsilon e^{-|z|^k})$ consists of all entire functions f satisfying

$$f = O(e^{N|z|^k})$$

for some positive integer N, that is the algebra of entire functions of finite order k.

4) Let $\Omega$ be an open set in $\mathbf{C}^n$. We associate to $\Omega$ the weight function $\delta_\Omega$ defined on $\mathbf{C}^n$ by

$$\delta_\Omega(s) = \mathrm{Min}\,(\delta_o(s),\ d(s, \complement\Omega)),$$

where $d(s, \complement\Omega)$ denotes the distance from s to the complement $\complement\Omega$ of $\Omega$. Functions in $\mathcal{O}(\delta_\Omega)$ are called holomorphic functions with polynomial growth on $\Omega$.

## 1.3.- Elementary properties

We consider in this section a non negative function $\delta$ on $\mathbf{C}^n$ satisfying condition W 2 (for instance a weight function). For s, s' in $\mathbf{C}^n$ such that $|s - s'| \leqslant \frac{1}{2}\delta(s)$, we have $|\delta(s') - \delta(s)| \leqslant \frac{1}{2}\delta(s)$ and

$$\frac{1}{2}\delta(s) \leqslant \delta(s') \leqslant \frac{3}{2}\delta(s).$$

Such an easy property will be often used.

Proposition 1.- Let $\delta$ be a non negative function on $\mathbf{C}^n$ satisfying condition W 2; then each $\partial/\partial z_j$ maps $\mathcal{O}(\delta)$ into $\mathcal{O}(\delta)$.

In other words, we have $\mathcal{O}(\delta) \subset \mathcal{C}_r(\delta)$ for every positive integer r.

Proof. Let f be a function of $\mathcal{O}(\delta)$; there exists a positive integer N such that $\delta^N|f|$ is uniformly bounded. We have to prove that $\partial f/\partial z_j$ satisfies a similar estimate. Let s be a point in $\{\delta > 0\}$; we consider the polydisc D with center at s and radius $r = \delta(s)/2\sqrt{n}$. Obviously D is contained in the ball B with center at s and radius $\frac{1}{2}\delta(s)$. Hence, for $\zeta$ in D, we have $\delta(\zeta) \geq \frac{1}{2}\delta(s)$ and D is contained in $\{\delta > 0\}$. Using Cauchy's formula, we get

$$f(s) = \frac{1}{(2\pi i)^n} \int_{|\zeta_1| = \dots = |\zeta_n| = r} \frac{f(\zeta)\, d\zeta_1 \dots d\zeta_n}{(\zeta_1 - s_1) \dots (\zeta_n - s_n)}$$

and

$$\frac{\partial f}{\partial z_j}(s) = \frac{1}{(2\pi i)^n} \int_{|\zeta_1| = \dots = |\zeta_n| = r} \frac{f(\zeta)\, d\zeta_1 \dots d\zeta_n}{(\zeta_1 - s_1) \dots (\zeta_j - s_j)^2 \dots (\zeta_n - s_n)} .$$

Hence

$$\frac{\partial f}{\partial z_j}(s) = \frac{1}{(2\pi)^n r} \int_0^{2\pi} \dots \int_0^{2\pi} f(s + r(e^{it_1}, \dots, e^{it_n}))\, e^{-it_j}\, dt_1 \dots dt_n .$$

If M is a uniform bound for $\delta^N |f|$, as

$$|f(\zeta)| \leq M\, \delta(\zeta)^{-N} \leq M\left(\frac{\delta(s)}{2}\right)^{-N}$$

for $\zeta$ in D, we get

$$\left|\frac{\partial f}{\partial z_j}(s)\right| \leq \frac{2\sqrt{n}}{\delta(s)} \left(\frac{\delta(s)}{2}\right)^{-N} M ,$$

that is

$$\delta(s)^{N+1} \left|\frac{\partial f}{\partial z_j}(s)\right| \leq \sqrt{n}\; 2^{N+1}\, M ,$$

and the result is proved, as N and M are independent of s.

Let $d\lambda$ denote the Lebesgue measure in $\mathbf{C}^n$, and p be a positive number.

Proposition 2.- Let $\delta$ be a weight function on $\mathbf{C}^n$; the algebra $\mathcal{O}(\delta)$ is the set of all holomorphic functions f on $\{\delta > 0\}$ satisfying

$$\int |f(s)|^p\, \delta^N(s)\, d\lambda(s) < +\infty ,$$

for some positive integer N.

Proof. We suppose first that f belongs to $\mathcal{O}(\delta)$. If M is a uniform bound for some $\delta^N |f|$, we get, assuming $\delta \leq 1$,

$$\int |f(s)|^p\, \delta^{N'}(s)\, d\lambda(s) \leq M^p \int \delta^{2n+2}(s)\, d\lambda(s),$$

for $N' \geq pN + 2n + 2$. As $\delta^{2n+2} = O(|z|^{-2n-2})$, the right side of the inequality is finite.

Conversely, if

$$\int |f(s)|^p \delta^N(s)\, d\lambda(s) < +\infty ,$$

for some positive integer N, and if f is holomorphic, we consider for s in $\{\delta>0\}$, the ball B with center at s and radius $\frac{1}{2}\delta(s)$. As $\delta(\zeta) \geq \frac{1}{2}\delta(s)$ for $\zeta$ in B, the ball B is contained in $\{\delta>0\}$ and from subharmonicity of $|f|^p$, we get

$$|f(s)|^p \leqslant \frac{1}{\mathrm{vol}(B)} \int_B |f(\zeta)|^p\, d\lambda(\zeta),$$

and

$$|f(s)|^p \delta^N(s) \leqslant \frac{1}{\mathrm{vol}(B)} \int_B |f(\zeta)|^p \delta^N(s)\, d\lambda(\zeta).$$

Upon replacing $\delta^N(s)$ by $\delta^N(\zeta)$ in the integrand, we only lose a multiple $2^N$ and the result follows from

$$1/\mathrm{vol}(B) = O(1/\delta^{2n}(s)).$$

1.4.- Regularization of weight functions

Any non negative function $\delta$ on $\mathbf{C}^n$ gives naturally rise to a weight function. We define first $\tilde{\delta}$ by

$$\tilde{\delta}(s) = \inf_{s' \in \mathbf{C}^n} (\delta(s') + |s'-s|).$$

Obviously $\tilde{\delta}$ satisfies condition W 2 and is smaller than $\delta$. Precisely $\tilde{\delta}$ is the largest function with such properties. In order to get a weight function, we only have to consider Min $(\delta_o, \tilde{\delta})$.

As algebras associated to equivalent functions are the same, all properties proved in Section 1.3 are valid when $\delta$ is only equivalent to a weight function. We give now a necessary and sufficient condition for this

Proposition 3.- Let $\delta$ be a non negative function on $\mathbf{C}^n$; in order that $\delta$ is equivalent to a weight function, it is necessary and sufficient that the following conditions are fulfilled :

H 1) Restrictions of polynomials belong to $\mathcal{C}(\delta)$.

H 2) There exist a positive integer N and $\varepsilon>0$ such that $|s-s'|<\varepsilon\delta^N(s)$ implies $\delta(s') \geq \varepsilon\delta^N(s)$.

Proof. Necessity has been proved in Section 1.1 for condition H 1, and in Section 1.3 for condition H 2 with $N=1$, $\varepsilon = \frac{1}{2}$.

We suppose now that conditions H 1, H 2 are fulfilled; as each $z_j$ belongs to

$\mathcal{C}(\delta)$, there exists a positive integer $N'$ such that each $|z_j|\delta^{N'}$ is uniformly bounded. As the constant 1 belongs to $\mathcal{C}(\delta)$, we get that $\delta$ is uniformly bounded. Hence we may assume that $N \geqslant N'$ and $\varepsilon\delta^N \leqslant \delta$. We take $\gamma = \varepsilon\delta^N$ and

$$\tilde{\gamma}(s) = \inf_{s \in \mathbf{C}^n} (\gamma(s') + |s'-s|).$$

As $\gamma \leqslant \varepsilon\delta^N$ and $|z|\delta^{N'}$ is uniformly bounded, $\tilde{\gamma}$ satisfies condition W1; hence $\tilde{\gamma}$ is a weight function. We also have $\tilde{\gamma} \leqslant \delta$. In the other direction, we use the following lemma to prove that $\tilde{\gamma}$ is equivalent to $\delta$.

Lemma 1.- Let $\delta$, $\delta'$ be two non negative functions on $\mathbf{C}^n$ such that $\delta' \geqslant \delta$. We assume that for some positive integer $N$ and some $\varepsilon > 0$, the relation $|s'-s| < \varepsilon\delta^N(s)$ implies $\delta(s') \geqslant \varepsilon\delta'^N(s)$; then $\tilde{\delta} \geqslant \varepsilon\delta'^N$.

We only have to use $\delta(s') + |s'-s| \geqslant |s'-s|$ if $|s'-s| \geqslant \varepsilon\delta^N(s)$ and $\delta(s') + |s'-s| \geqslant \delta(s')$ if $|s'-s| < \varepsilon\delta^N(s)$, to get

$$\delta(s') + |s'-s| \geqslant \varepsilon\delta'^N(s)$$

in all cases. Taking the infimum on $s'$, we have $\tilde{\delta} \geqslant \varepsilon\delta'^N$.

As an example, every non negative function $\delta$ on $\mathbf{C}^n$ satisfying W1 or H1 and such that

$$|\delta(s) - \delta(s')| \leqslant C|s-s'|^{\alpha}$$

with $C \geqslant 0$, $\alpha > 0$, is equivalent to a weight function : if $N$, $\varepsilon$ are chosen such that $N \geqslant 1/\alpha$, $C\varepsilon^{\alpha} \leqslant \frac{1}{2}$ and if $|s-s'| \leqslant \varepsilon\delta^N(s)$, we have

$$\delta(s') \geqslant \delta(s) - C\varepsilon^{\alpha}\delta(s) \geqslant \tfrac{1}{2}\delta(s)$$

and condition H2 follows.

We shall see that every weight function $\delta$ is equivalent to a weight function which is $\mathcal{C}^{\infty}$ on $\{\delta > 0\}$. More precisely

Proposition 4.- Let $\delta$ be a weight function on $\mathbf{C}^n$ and $\varepsilon$ be a strictly positive constant. There exists a function $\delta'$, which is $\mathcal{C}^{\infty}$ on $\{\delta > 0\}$, such that

$$(1-\varepsilon)\delta \leqslant \delta' \leqslant (1+\varepsilon)\delta$$

and

$$|D\delta'| = O((\delta\varepsilon)^{1-r})$$

for every derivative $D$ of order $r \geqslant 1$.

Proof. We may assume that $\delta \leqslant 1$, $\varepsilon < 1$. For every non negative integer $p$, let $S_p$ denote the set of all points $s$ in $\mathbf{C}^n$ such that $\delta(s) \geqslant (1-\varepsilon)^p$. We consider a non negative $\mathcal{C}^{\infty}$ function $\varphi$ on $\mathbf{C}^n$, with support in the unit ball $\{|z| \leqslant 1\}$, such that

$$\int_{\mathbf{C}^n} \varphi(\zeta)\, d\lambda(\zeta) = 1,$$

and we define, for every p, a function $\varphi_p$ by

$$\varphi_p(s) = (\varepsilon(1-\varepsilon)^p)^{-2n}\, \varphi\left(\frac{s}{\varepsilon(1-\varepsilon)^p}\right).$$

Let also $\chi_p$ be the characteristic function of $S_p$ and

$$\delta' = \sum_{p\geq 0} \varepsilon(1-\varepsilon)^p\, \chi_p * \varphi_p .$$

As $\int_{\mathbf{C}^n} \varphi_p(\zeta)\, d\lambda(\zeta) = 1$ for every p, we have $|\chi_p * \varphi_p| \leq 1$ and the series is uniformly convergent.

We shall first estimate $\delta'$; note that if s belongs to $S_p$ and satisfies

$$|s - s'| \leq \varepsilon(1-\varepsilon)^p,$$

then s' belongs to $S_{p+1}$, because

$$\delta(s') \geq (1-\varepsilon)^p - \varepsilon(1-\varepsilon)^p = (1-\varepsilon)^{p+1}.$$

Hence $\chi_q * \varphi_q(s) = 1$ for $q \geq p+1$, as $|s-s'| \leq \varepsilon(1-\varepsilon)^p$ for every point s' in the support of $\varphi_q$, and

$$\delta'(s) \geq \sum_{q\geq p+1} \varepsilon(1-\varepsilon)^q = (1-\varepsilon)^{p+1}.$$

On the other hand, if s does not belong to $S_{p-1}$, we have $\chi_q * \varphi_q(s) = 0$ for $q \leq p-2$ by a similar argument, and

$$\delta'(s) \leq \sum_{q\geq p-1} \varepsilon(1-\varepsilon)^q = (1-\varepsilon)^{p-1}.$$

Thus

$$(1-\varepsilon)^p \leq \delta(s) < (1-\varepsilon)^{p-1}$$

implies

$$(1-\varepsilon)^{p+1} \leq \delta'(s) < (1-\varepsilon)^{p-1}.$$

As the property is valid for every p, we get

$$(1-\varepsilon)^2\, \delta(s) \leq \delta'(s) \leq \frac{\delta(s)}{1-\varepsilon},$$

and the good estimate follows, when replacing $\varepsilon$ by $\varepsilon/2$.

Let D be a derivative of order $r \geq 1$. We have

$$D\delta' = \sum_{p\geq 0} \varepsilon(1-\varepsilon)^p\, \chi_p * D\varphi_p ,$$

and the series is locally uniformly convergent, because if

$$(1-\varepsilon)^p \leqslant \delta(s) < (1-\varepsilon)^{p-1},$$

only $D\varphi_{p-1}$ and $D\varphi_p$ do not vanish at $s$. Moreover

$$|D\delta'(s)| \leqslant \varepsilon(1-\varepsilon)^{p-1}\|D\varphi_{p-1}\|_\infty + \varepsilon(1-\varepsilon)^p\|D\varphi_p\|_\infty .$$

As

$$\|D\varphi_q\|_\infty = (\varepsilon(1-\varepsilon)^q)^{-r}\|D\varphi\|_\infty,$$

we get

$$|D\delta'(s)| = O(\varepsilon^{1-r}(1-\varepsilon)^{p(1-r)}),$$

and finally

$$|D\delta'(s)| = O((\varepsilon\delta(s))^{1-r}).$$

Thus the proof is complete.

## 1.5.- Inductive limits

In this section we consider algebras of holomorphic functions which are inductive limits of algebras introduced in Section 1.1.

Let $\Delta$ be a set of non negative functions on $\mathbf{C}^n$; we assume that $\Delta$ is directed in the following sense : for every $\delta', \delta'' \in \Delta$, there exists some $\delta \in \Delta$ such that $\varepsilon\delta^N \leqslant \delta'$, $\varepsilon\delta^N \leqslant \delta''$ for some positive integer N and some $\varepsilon > 0$. By considering, for $\delta, \delta'$ in $\Delta$ such that $\varepsilon\delta^N \leqslant \delta'$ for some N, $\varepsilon$, the restriction mapping

$$\mathcal{C}(\delta') \to \mathcal{C}(\delta),$$

we define an inductive system. The direct limit of this system is an algebra; it is denoted by $\mathcal{C}(\Delta)$. Elements of $\mathcal{C}(\Delta)$ are obtained by identifying in $\bigcup_{\delta\in\Delta}\mathcal{C}(\delta)$ functions which are equal on some $\{\delta>0\}$.

If each $\{\delta>0\}$ is assumed to be open, the direct limit of the system $\mathcal{O}(\delta)$ is a subalgebra of $\mathcal{C}(\Delta)$; it is denoted by $\mathcal{O}(\Delta)$.

We discuss a few examples of algebras $\mathcal{O}(\Delta)$, where $\Delta$ is a directed set of weight functions or can be replaced by such a directed set.

1) Let $\Omega$ be an open set in $\mathbf{C}^n$. For every convex increasing mapping of $[0,\infty[$ into $[0,\infty[$, we set

$$\delta_{\Omega,\varphi} = \exp(-\varphi(-\log\delta_\Omega)).$$

Then $\Delta_\Omega$ denotes the set of all functions $\delta_{\Omega,\varphi}$ which are weight functions. It is easily seen that $\Delta_\Omega$ is a directed set because the function $\operatorname{Min}(\delta_{\Omega,\varphi}, \delta_{\Omega,\varphi'})$ is associated to $\mathrm{Max}(\varphi, \varphi')$ and belongs to $\Delta_\Omega$.

Proposition 4.- <u>For every open set</u> $\Omega$ <u>in</u> $\mathbf{C}^n$, <u>the algebra</u> $\mathcal{O}(\Delta_\Omega)$ <u>is the algebra</u> $\mathcal{O}(\Omega)$ <u>of all holomorphic functions on</u> $\Omega$ .

It is an easy consequence of the following

Lemma 2.- <u>For every increasing mapping</u> $\psi$ <u>from</u> $[0,\infty[$ <u>to</u> $[0,\infty[$ , <u>there exists a convex increasing mapping</u> $\varphi$ <u>from</u> $[0,\infty[$ <u>to</u> $[0,\infty[$ <u>such that</u> $\psi \leqslant \varphi$ <u>and</u>

$$x \mapsto f(x) = \exp(-\varphi(-\log x))$$

<u>is Lipschitz with constant</u> 1 <u>on</u> $]0, 1]$ .

<u>Proof of lemma 2</u>. We consider the sequence $c_n = \psi(n+1)$ and define $g$ by induction on $[0, n]$ as a convex increasing majorant of $\psi$ such that

$$g'(y) \leqslant e^{g(y)}$$

for every $y$ in $[0, n]$ , where $g'(y)$ is the left or right derivative at $y$, and

$$g(n) \geqslant c_n.$$

We define $g$ on $\{0\}$ by $g(0) = c_o$ and, if $g$ has been already defined on $[0, n]$ , we consider the unique solution defined on a right neighbourhood of $n$ of

$$\frac{dh}{dy} = e^h$$

such that $h(n) = g(n)$. Thus

$$y = n + e^{-g(n)} - e^{-h(y)},$$

and $h$ tends to infinity at a point $c$ of $]n, n+1[$. Obviously $h$ is convex increasing on $[n, c[$ and, as $h'(n) = e^{g(n)} \geqslant g'(n)$, if we extend $g$ by $h$, we obtain a convex increasing function on $[0, c[$. By replacing $h$ on $[c', +\infty[$ by the affine function which is tangent to $h$ at a point $c'$ of $]n, c[$, we obtain a convex increasing function on $[0, +\infty[$ and we may assume that its value at $n+1$ is larger than $c_{n+1}$. We take this function as $g$ on $[0, n+1]$ . As $g(n) \geqslant \psi(n+1)$, and $g, \psi$ are both increasing, we have $g(y) \geqslant \psi(y)$ for every $y$ in $]n, n+1]$ and the induction is complete.

Let $\varphi$ be defined by

$$\varphi(y) = 1 + y + g(y).$$

Obviously $\varphi$ is a convex increasing majorant of $\psi$ . Moreover

$$f'(x) = e^{y - \varphi(y)} \varphi'(y) = e^{-1 - g(y)}(1 + g'(y)),$$

and as $e^{-g(y)} g'(y) \leqslant 1$, we get $|f'(x)| \leqslant 1$. Hence $f$ is Lipschitz with constant 1.

Proof of Proposition 4. Let $\psi$ be a convex increasing mapping of $[0,\infty[$ into $[0,\infty[$ and $\varphi$ as in Lemma 2. As $\delta_\Omega$ is Lipschitz with constant 1, $\delta_{\Omega,\varphi}$ is also Lipschitz with constant 1 and therefore satisfies condition W 2. If we assume that $\psi(y) \geqslant y$, we get $\delta_{\Omega,\varphi} \leqslant \delta_\Omega$ and $\delta_{\Omega,\varphi}$ also satisfies condition W 1. We only have to prove that every f in $\mathcal{O}(\Omega)$ satisfies

$$\exp(-\psi(-\log \delta_\Omega))\,|f| \leqslant 1$$

or

$$\log|f| \leqslant \psi(-\log \delta_\Omega),$$

for some increasing mapping of $[0,\infty[$ into $[0,\infty[$. As the set $\{\delta_\Omega \geqslant e^{-y}\}$ is compact in $\Omega$ for every $y \in [0,\infty[$, we can take

$$\psi(y) = \sup_{\delta_\Omega(s) \geqslant e^{-y}} \log^+|f(s)|.$$

2) Let K be a compact set in $\mathbf{C}^n$. If $\Delta_K$ denotes the set of all weight functions $\delta_\Omega$, where $\Omega$ is an open neighbourhood of K, the algebra $\mathcal{O}(\Delta_K)$ is the algebra $\mathcal{O}(K)$ of germs of functions which are holomorphic on a neighbourhood of K.

3) The algebra of entire functions of finite order on $\mathbf{C}^n$ is the direct limit of the algebras $\mathcal{O}(e^{-|z|^k})$, where k varies in $\mathbf{N}$.

Let us more generally consider a non bounded convex increasing mapping of $[0,\infty[$ into $[0,\infty[$, called $\varphi$.

We denote by $\Lambda_\varphi$ the set of all functions $\lambda_c$ where

$$\lambda_c = \exp(-\varphi(-\log c\,\delta_0))$$

and c is a positive constant. Clearly $\mathcal{O}(\Lambda_\varphi)$ is the algebra of all entire functions f on $\mathbf{C}^n$ such that

$$|f| = O(e^{C\,\varphi(-\log c\,\delta_0)}),$$

for some $c>0$, $C \geqslant 0$. As $\log|z| \leqslant -\log\delta_0$ and $-\log(\frac{1}{2}\delta_0) \leqslant \log|z|$ for $|z| \geqslant 1$, it is easily seen that $\mathcal{O}(\Lambda_\varphi)$ is the algebra of all entire functions f such that

$$|f| = O(e^{C\,\varphi(-\log|z|+C')})$$

for some $C' \geqslant 0$, $C \geqslant 0$.

We can find constants $\alpha>0$, $M \geqslant 0$, such that $\varphi(x) \geqslant \alpha x - M$ for all $x \in [0,\infty[$. Therefore $\lambda_c = O(|z|^{-\alpha})$ at infinity and is equivalent to some function satisfying W 1.

Maybe $\lambda_c$ is not equivalent to some weight function. We shall see, however, that $\mathcal{O}(\Lambda_\varphi)$ is equal to $\mathcal{O}((\tilde{\lambda}_c))$. It suffices to show that $\tilde{\lambda}_c$ is larger than some function equivalent to $\lambda_{c/2}$. We use Lemma 1 of Section 1.4 with $\delta = \lambda_c$ and $\delta' = \lambda_{c/2}$

and remark that $|s-s'| \leqslant 1/2\ \delta_o(s)$ implies $\delta_o(s') \geqslant 1/2\ \delta_o(s)$ and

$$\lambda_c(s') \geqslant \lambda_{c/2}(s).$$

To complete the proof, we remember that $\delta_o$ is larger than some function equivalent to $\lambda_c$.

When $\Omega$ is an open set in $\mathbf{C}^n$ and $\varphi$ as above, we denote by $\Lambda_{\Omega,\varphi}$ the set of all functions $\lambda_c$ where

$$\lambda_c = \exp(-\varphi(-\log c\ \delta_\Omega)).$$

Now $\mathcal{O}(\Lambda_{\Omega,\varphi})$ is the algebra of all holomorphic functions f on $\Omega$ such that

$$|f| = o(e^{C\ \varphi(-\log c\ \delta_\Omega)})$$

for some $c > 0$, $C \geqslant 0$. We similarly show that $\lambda_c$ is equivalent to some function satisfying W 1 and that $\mathcal{O}(\Lambda_{\Omega,\varphi}) = \mathcal{O}((\tilde{\lambda}_c))$.

Thus $\mathcal{O}(\Lambda_\varphi)$ or $\mathcal{O}(\Lambda_{\Omega,\varphi})$ are equal to algebras $\mathcal{O}(\Delta)$, where $\Delta$ is a directed set of weight functions.

Proposition 1 of Section 1.3 can be generalized in the following way:

Proposition 5.- <u>Assume that every</u> $\delta \in \Delta$ <u>satisfies condition</u> W 2; <u>then</u>, $\partial/\partial z_j$ <u>maps</u> $\mathcal{O}(\Delta)$ <u>into</u> $\mathcal{O}(\Delta)$.

Let $\Delta$, $\Delta'$ be two directed sets of weight functions on $\mathbf{C}^n$. We suppose that for every $\delta' \in \Delta'$ there exists $\delta \in \Delta$ such that $\delta' \geqslant \varepsilon\,\delta^N$ for some positive integer N and some $\varepsilon > 0$. The set of restriction mappings $\mathcal{C}(\delta') \to \mathcal{C}(\delta)$ gives rise to an homomorphism $\mathcal{C}(\Delta') \to \mathcal{C}(\Delta)$. We similarly have an homomorphism $\mathcal{O}(\Delta') \to \mathcal{O}(\Delta)$; if we also suppose that for every $\delta' \in \Delta'$ the set $\{\delta' > 0\}$ is connected, and that no $\delta \in \Delta$ is identically zero, it can be easily proved that this last morphism is injective: in such a case, $\mathcal{O}(\Delta')$ can be considered as a subalgebra of $\mathcal{O}(\Delta)$.

<u>Notes</u>.

Algebras $\mathcal{C}(\delta)$, $\mathcal{C}_r(\delta)$ and $\mathcal{O}(\delta)$, and conditions W 1, W 2 have been introduced by L. Waelbroeck (1) (2). Propositions 1 and 2 are standard; for a different proof of Proposition 2, see I. Cnop (2). Conditions H 1, H 2 have been considered by L. Hörmander (3), and J.J. Kelleher and B.A. Taylor (1) : they denote by $A(\varphi)$ the algebra $\mathcal{O}(\delta)$ when $\delta = e^{-\varphi}$. Proposition 4 is stronger than a similar result proved by L. Waelbroeck (1) and uses a method taken from a paper of the author (2). Algebras of holomorphic functions of exponential type or of finite order are classical: see C.O. Kiselman (1) and A. Martineau (1). Algebras $\mathcal{O}(\Lambda_\varphi)$ have been considered by L. Rubel and B.A. Taylor (1) and denoted by $E(\lambda)$ where $\lambda$ is the function $x \mapsto \varphi(\log x)$.

# CHAPTER II

## BOUNDEDNESS AND POLYNORMED VECTOR SPACES

We introduce a structure on the algebras of holomorphic functions with growth we have considered in Chapter I. This is not a topology but a boundedness. We define first polynormed vector spaces as vector spaces equipped with a suitable covering by pseudonormed spaces. We introduce then bounded sets, convergent and Cauchy sequences. This leads to the definition of Hausdorff and complete polynormed vector spaces.

### 2.1.- Polynormed vector spaces

We consider along this section a non negative function $\delta$ on $\mathbf{C}^n$ which is assumed to be bounded; we have defined in Chapter I the algebra $\mathcal{T}(\delta)$ of $\delta$-tempered complex-valued functions on $\mathbf{C}^n$. In the general case, there is no way to equip $\mathcal{T}(\delta)$ with a norm. However $\mathcal{T}(\delta)$ naturally carries the family of pseudonorms

$$f \mapsto \sup_{\delta(s)>0} |f(s)|\, \delta^N(s),$$

where N ranges over $\mathbf{Z}$.

It is easily seen that these pseudonorms are equivalent if and only if $\delta$ and $1/\delta$ are both bounded on the set $S = \{\delta > 0\}$ ; this means that $\delta$ is equivalent to the characteristic function of S. In such a case $\mathcal{T}(\delta)$ is the Banach algebra of all bounded complex-valued functions on S. We note that a weight function $\delta$ which does not identically vanish is never a characteristic function.

Returning to the general case, we have

$$\mathcal{T}(\delta) = \bigcup_{N\in\mathbf{Z}} {}_N\mathcal{T}(\delta),$$

where ${}_N\mathcal{C}(\delta)$ is the space of all complex-valued functions f on $\{\delta>0\}$ such that $|f|\delta^N$ is bounded, equipped with the norm

$$f \mapsto \sup_{\delta(s)>0} |f(s)|\,\delta^N(s).$$

Each ${}_N\mathcal{C}(\delta)$ is a Banach space; as $\delta$ is bounded, ${}_N\mathcal{C}(\delta)$ is contained in ${}_{N+1}\mathcal{C}(\delta)$ and the identity mapping

$${}_N\mathcal{C}(\delta) \longrightarrow {}_{N+1}\mathcal{C}(\delta)$$

is continuous. It would be possible to consider on $\mathcal{C}(\delta)$ the direct limit locally convex topology of the sequence ${}_N\mathcal{C}(\delta)$, but such a topology is not easy to handle. We therefore introduce structures which are closer to the estimates leading to the definition of $\mathcal{C}(\delta)$. Roughly speaking we do not take the limit but prefer work with the system itself.

By definition a polynormed vector space is a complex vector space E equipped with a covering $({}_iE)_{i\in I}$ by pseudonormed (*) vector spaces such that I is a directed ordered set and, for every $i \leq j$, the identity mapping is continuous from ${}_iE$ into ${}_jE$.

Let $(E, ({}_iE)_{i\in I})$ and $(F, ({}_jF)_{j\in J})$ be polynormed vector spaces. We say that a linear mapping u of E into F is bounded if for every $i\in I$ there exists $j\in J$ such that u is a continuous mapping of ${}_iE$ into ${}_jF$.

List a few examples:

1) The vector space $\mathcal{C}(\delta)$ equipped with the covering $({}_N\mathcal{C}(\delta))_{N\in\mathbf{Z}}$ considered in the beginning of this section is a polynormed vector space.

2) When the set $\{\delta>0\}$ is supposed to be open, we can do the same for $\mathcal{O}(\delta)$ by considering for every $N\in\mathbf{Z}$ the vector space

$${}_N\mathcal{O}(\delta) = {}_N\mathcal{C}(\delta)\cap\mathcal{O}(\delta)$$

of all functions in ${}_N\mathcal{C}(\delta)$ which are holomorphic, equipped with the norm induced by ${}_N\mathcal{C}(\delta)$. The couple $(\mathcal{O}(\delta), ({}_N\mathcal{O}(\delta))_{N\in\mathbf{Z}})$ defines a polynormed vector space. Proposition 1 of Chapter I may be completed as following:

If $\delta$ is a non negative function on $\mathbf{C}^n$ satisfying condition W2, each $\partial/\partial z_j$ is a bounded linear mapping of $\mathcal{O}(\delta)$ into $\mathcal{O}(\delta)$.

3) We may also regard $\mathcal{C}_r(\delta)$ as a polynormed vector space; for every $N\in\mathbf{Z}$, we denote by ${}_N\mathcal{C}_r(\delta)$ the vector space of all functions f in $\mathcal{C}_r(\delta)$ such that $\delta^{N-s}|Df|$ is bounded for every derivative D of order $s\leq r$, equipped with the norm

$$f \mapsto \operatorname*{Max}_{|\alpha|\leq r}\Big(\sup_{\delta(s)>0} \delta^{N-|\alpha|}(s)\,|D^\alpha f(s)|\Big),$$

where $\alpha$ is a multi-index with length $|\alpha|$ and $D^\alpha$ the derivative associated to $\alpha$. The covering $({}_N\mathcal{C}_r(\delta))_{N\in\mathbf{Z}}$ satisfies the required conditions.

If $\delta'$ is another bounded non negative function on $\mathbf{C}^n$ such that $\delta' \geqslant \varepsilon\delta^N$ for some positive number $\varepsilon$ and some positive integer N, it is clear that the restriction mapping $\mathcal{C}(\delta')\to\mathcal{C}(\delta)$ (resp. $\mathcal{O}(\delta')\to\mathcal{O}(\delta)$, $\mathcal{C}_r(\delta')\to\mathcal{C}_r(\delta)$ if $\{\delta>0\}$, $\{\delta'>0\}$ are open) considered in Section 1.1 is a bounded linear mapping.

4) When $\Delta$ is a directed set of bounded non negative functions on $\mathbf{C}^n$, we equip $\mathcal{C}(\Delta)$ (resp. $\mathcal{O}(\Delta)$ when each $\{\delta>0\}$ is open) with a covering which makes it into a polynormed vector space; we consider on couples $(\delta, N)\in\Delta\times\mathbf{Z}$ the relation $(\delta, N)\leqslant(\delta', N')$ defined by

" $\delta'^N$ is bounded by a positive multiple of $\delta^N$ ",

and for each $(\delta, N)$, we take the pseudonormed vector space which is the image of ${}_N\mathcal{C}(\delta)$ (resp. ${}_N\mathcal{O}(\delta)$) in $\mathcal{C}(\Delta)$.

5) Let $\underline{E} = (E, ({}_iE)_{i\in I})$ be a polynormed vector space. We introduce a new polynormed vector space $\mathcal{C}(\delta;\underline{E})$ as the union, where i varies in I and N in $\mathbf{Z}$, of the pseudonormed vector spaces ${}_N\mathcal{C}(\delta; E_i)$ defined as following: ${}_N\mathcal{C}(\delta; E_i)$ consists of all functions f defined on $\{\delta>0\}$ and with values in ${}_iE$ such that $\delta^N(s)\, f(s)$ is uniformly bounded for $\delta(s)>0$.

When $\{\delta>0\}$ is open, we also define the subspaces ${}_N\mathcal{C}_r(\delta; E_i)$ (resp. ${}_N\mathcal{C}\mathcal{C}_r(\delta; E_i)$, ${}_N\mathcal{O}(\delta; E_i)$) of all functions in ${}_N\mathcal{C}(\delta; E_i)$ which are r-times differentiable (resp. r-times continuously differentiable, holomorphic). Taking the union when N, i vary, we obtain polynormed vector spaces denoted by $\mathcal{C}_r(\delta; E)$, $\mathcal{C}\mathcal{C}_r(\delta; E)$, $\mathcal{O}(\delta; E)$.

## 2.2.- Convex bounded structures

We may consider the category, the objects of which are pseudonormed vector spaces and morphisms are bounded linear mappings. The product of polynormed vector spaces $(E, ({}_iE)_{i\in I})$ and $(F, ({}_jF)_{j\in J})$ is defined as the product space $E\times F$ equipped with the covering $({}_iE\times{}_jF)_{(i,j)\in I\times J}$. When F is a vector subspace of a polynormed vector space $(E, ({}_iE)_{i\in I})$, we consider on F the induced covering $(F\cap{}_iE)_{i\in I}$, where $F\cap{}_iE$ is equipped with the pseudonorm of ${}_iE$.

Two polynormed vector spaces $(E, ({}_iE)_{i\in I})$, $(E, ({}_jF)_{j\in J})$ are isomorphic above E if the identity mapping of E is a morphism in both directions: for every $i\in I$, there exists $j\in J$ such that ${}_iE$ is contained in ${}_jF$ and has a finer pseudonorm, and inversely; if we replace $\mathbf{Z}$ by $\mathbf{N}$ when defining $\mathcal{C}(\delta)$ or $\mathcal{O}(\delta)$, we obtain isomorphic polynormed spaces. A class of polynormed vector spaces which are isomorphic above E is called a convex boundedness on E.

Let $(E, ({}_iE)_{i\in I})$ be a polynormed vector space. A subset B of E is said to be bounded if B is contained in an homothetic of the unit ball of some ${}_iE$. It is easily seen that a linear mapping u of a polynormed vector space into another is bounded if and only if the image by u of every bounded set is bounded. A convex boundedness on a vector space E is then uniquely determined by the set $\mathfrak{B}$ of all bounded subsets. We note first that $\mathfrak{B}$ has the following properties:

a) $B_1 \cup B_2 \in \mathfrak{B}$ if $B_1 \in \mathfrak{B}$, $B_2 \in \mathfrak{B}$.

b) $B' \in \mathfrak{B}$ if $B' \subset B$, $B \in \mathfrak{B}$.

c) E is the union of $\mathfrak{B}$.

In general a set $\mathfrak{B}$ of subsets of a given set E satisfying properties a), b), c) is called a <u>boundedness</u> or a <u>bounded structure</u> on E. The elements of $\mathfrak{B}$ are called the bounded sets of the boundedness. If $\mathfrak{B}$ is the boundedness of a polynormed vector space E, we also have:

CB) Every bounded set is contained in an absolutely convex bounded set.

Conversely, if $\mathfrak{B}$ is a boundedness on a vector space E satisfying condition CB), we may consider the ordered set $\mathcal{E}$ of all absolutely convex bounded sets of $\mathfrak{B}$ and for every $B \in \mathcal{E}$, the vector space $E_B$ spanned by B, equipped with the Minkowski functional

$$x \mapsto \|x\|_B = \inf_{\substack{x \in \lambda B \\ \lambda > 0}} \lambda$$

of B.

We thus define a polynormed vector space $(E, (E_B)_{B \in \mathcal{E}})$ and it is easily checked that the boundedness of such a polynormed vector space is $\mathfrak{B}$. Hence convex boundednesses on a vector space E are exactly boundednesses on E satisfying condition CB).

Note that a subset B of $\mathcal{C}(\delta)$ (resp. $\mathcal{O}(\delta)$) is bounded if and only if there exist a positive integer N and a positive number M such that B is contained in the set of all complex-valued functions on $\{\delta > 0\}$ such that

$$|f|\,\delta^N \leqslant M.$$

Convex boundednesses are often used as tools in the study of topological vector spaces. For instance, if E is a locally convex topological vector space, the Mackey boundedness of E is the set of all subsets B of E such that for every neighbourhood U of the origin there exists some $\varepsilon > 0$ with $\varepsilon B \subset U$.

Proposition 4 of Chapter I is completed as following:

Proposition 1.- <u>Let $\Omega$ an open set in $\mathbf{C}^n$; the boundedness of $\mathcal{O}(\Delta_\Omega)$ is also the Mackey boundedness of $\mathcal{O}(\Omega)$, when equipped with the compact open topology</u>.

<u>Proof</u>. A subset B of $\mathcal{O}(\Omega)$ is bounded for the compact open topology if and only if for every compact set K in $\Omega$ there exists a constant $M_K$ which is a uniform bound on

K for all functions in B. From this follows that every bounded set in $\mathcal{O}(\Delta_\Omega)$ is bounded in $\mathcal{O}(\Omega)$. Conversely, if B is bounded in $\mathcal{O}(\Omega)$, taking

$$\psi(y) = \sup_{\delta_\Omega(s) \geqslant e^{-y}} (\sup_{f \in B} \log^+|f(s)|),$$

and a majorant $\varphi$ of $\psi$ such that $\delta_{\Omega,\varphi}$ is a weight function, B is bounded in $\mathcal{O}(\delta_{\Omega,\varphi})$ and thereby in $\mathcal{O}(\Delta_\Omega)$.

Note that the boundedness of a polynormed vector space $\underline{E}$ is also determined by the set of all pseudonormed vector subspaces N of E such that the identity $N \to E$ is a bounded linear mapping. Such pseudonormed vector spaces N will be called pseudonormed vector spaces of the definition of $\underline{E}$.

## 2.3.- Completeness

Let $\underline{E} = (E, ({}_iE)_{i \in I})$ be a polynormed vector space, we say that it is Hausdorff if each ${}_iE$ is a normed space; this means that there exists no bounded linear subspace except $\{0\}$. We say that it is complete if, up to isomorphism above E, each ${}_iE$ is a Banach space; it is equivalent to ask that every bounded subset of E is contained in an absolutely convex bounded subset B such that $E_B$ is a Banach space.

A sequence $(x_p)$ in E is said to converge to $x \in E$ (resp. to be a Cauchy sequence) if there exists some ${}_iE$ such that the property holds in ${}_iE$. This is equivalent to the existence of a bounded subset B and a sequence $\varepsilon_p$ tending to zero in $\mathbf{C}$ such that

$$x_p - x \in \varepsilon_p B$$

(resp.

$$x_{p+q} - x_p \in \varepsilon_p B$$

for $q \geqslant 0$).

Note that every convergent sequence has a unique limit in a polynormed vector space if and only if it is Hausdorff. If $\underline{E}$ is Hausdorff, a necessary and sufficient condition for completeness is that for every $i \in I$ we can find some $j \in I$ such that every Cauchy sequence in ${}_iE$ is convergent in ${}_jE$. To prove the sufficiency, choose for every $i \in I$ some $j \in I$ with the above property and such that ${}_iE$ is continuously mapped into ${}_jE$. We have morphisms

$${}_iE \to {}_i\hat{E} \to {}_jE.$$

Taking the image ${}_iF$ of ${}_i\hat{E}$ in ${}_jE$, we also have morphisms

$${}_iE \to {}_iF \to {}_jE.$$

Obviously, each ${}_iF$ is complete and $({}_iF)_{i \in I}$ defines a polynormed vector space which is isomorphic to E.

Let $(x_p)$ be a sequence of E and $(\lambda_p)$ a sequence of positive numbers. We write $x_p = O(\lambda_p)$ (resp. $x_n = o(\lambda_p)$) if there exists a bounded sequence $(y_p)$ (resp. a sequence $(y_p)$ tending to zero) such that $x_p = \lambda_p y_p$.

We consider a bounded non negative function $\delta$ on $\mathbf{C}^n$. As each ${}_N\mathcal{C}(\delta)$ is a Banach space, $\mathcal{C}(\delta)$ is complete. Further, as easily shown:

Proposition 2.- _If $\delta$ is lower semi-continuous, convergence in $\mathcal{C}(\delta)$ implies compact convergence in $\{\delta>0\}$ and $\mathcal{O}(\delta)$, $\mathcal{C}_r(\delta)$ are complete._

$\mathcal{O}(\Delta)$ is not always complete when $\mathcal{O}(\delta)$ is complete for every $\delta\in\Delta$. For instance, if $\omega_p$ is the open disc in the complex plane with center at $p\in\mathbf{N}$ and radius $\frac{1}{2}$, and $\Omega_p = \bigcup_{q\geqslant p}\omega_q$, as $\delta_{\Omega_p}$ is a weight function, we know that $\mathcal{O}(\delta_{\Omega_p})$ is complete. However, if $\Delta$ is the set of all $\delta_{\Omega_p}$, it is easily checked that $\mathcal{O}(\Delta)$ is not Hausdorff: choose a sequence $\varepsilon_p$ of positive numbers tending to zero and define f in $\mathcal{O}(\delta_{\Omega_0})$ by $f(s) = \varepsilon_p$ for $s\in\omega_p$; replacing f by zero on $\omega_1\cup\ldots\cup\omega_p$ does not change the class in $\mathcal{O}(\Delta)$ and therefore f tends to zero in $\mathcal{O}(\Delta)$.

An example of complete $\mathcal{O}(\Delta)$ is given by

Proposition 3.- _Let $\Delta$ be a directed set of bounded lower semi-continuous non negative functions on $\mathbf{C}^n$; if $\{\delta>0\}$ has finitely many connected components for every $\delta\in\Delta$, then $\mathcal{O}(\Delta)$ is complete._

We only have to prove that $\mathcal{O}(\Delta)$ is hausdorff: each ${}_N\mathcal{O}(\delta)$ is a Banach space; if the image of ${}_N\mathcal{O}(\delta)$ in $\mathcal{O}(\Delta)$ is a normed space, this is a Banach space. By absurd, assume that f tends to zero in $\mathcal{O}(\Delta)$; this means that there exist $\delta_1\in\Delta$ and a sequence $(f_p)$ tending to zero in $\mathcal{O}(\delta_1)$ such that $f_p$ represents f for every p. Let $\omega$ be the union of all connected components of $\{\delta_1>0\}$ which intersect $\{\delta>0\}$ for every $\delta\in\Delta$. We can choose some $\delta_2\in\Delta$ such that $\varepsilon\delta_2^N\leqslant\delta_1$ for some positive integer N and some positive number $\varepsilon$ and that $\{\delta_2>0\}$ is contained in $\omega$. As $f_p$ and $f_q$ coincide on some $\{\delta>0\}$ and are holomorphic on $\{\delta_1>0\}$, they coincide on $\omega$ and on $\{\delta_2>0\}$. Then $(f_p)$ is a constant sequence tending to zero in $\mathcal{O}(\delta_2)$; thus $f_p = 0$ on $\{\delta_2>0\}$ and $f = 0$.

We associate to every polynormed vector space $\underline{E} = (E, ({}_iE)_{i\in I})$ a complete polynormed vector space as follows: for each $i\in I$, we denote by ${}_i\hat{E}$ the Banach space associated to ${}_iE$ and by ${}_iF$ the image of ${}_i\hat{E}$ in the direct limit F of the system ${}_i\hat{E}$. Obviously $(F, ({}_iF)_{i\in I})$ is complete; it is denoted by $\hat{\underline{E}}$.

2.4.- Closure and density

Let $(E, ({}_iE)_{i\in I})$ be a polynormed vector space. If $A$ is a subset of $E$, we denote by $\bar{A}$ the set of limits in $E$ of elements of $A$. We say that $A$ is closed if $\bar{A} = A$. It is important to note that $\bar{A}$ is not necessarily closed; the closure of $A$, defined as the smallest closed subset containing $A$, may require transfinitely many operations.

Now consider a vector subspace $F$ of $E$. We remark that, by definition of limits in $E$, we have

$$\bar{F} = \bigcup_{i\in I} \mathrm{Cl}_{{}_iE}(F \cap {}_iE),$$

where $\mathrm{Cl}_{{}_iE}(F \cap {}_iE)$ is the closure in ${}_iE$ of $F \cap {}_iE$. The covering $(\mathrm{Cl}_{{}_iE}(F \cap {}_iE))_{i\in I}$ enables us to consider $\bar{F}$ as a polynormed vector space. We always have morphisms

$$F \to \bar{F} \to E,$$

but $\bar{F}$ is not in general isomorphic to a vector subspace of $E$.

A vector subspace $F$ of $E$ is called dense in $E$ when the polynormed vector spaces $\bar{F}$ and $E$ are isomorphic. This means that for each $i\in I$, there exists $j \in J$ such that every element of ${}_iE$ is a limit of elements of $F$ according to the pseudonorm of ${}_jF$. This implies in particular that elements of $E$ are limits of elements of $F$.

If $F$ is a vector subspace of a complete polynormed vector space $E$, it is easily seen that $\bar{F} = \hat{F}$ ; therefore $\bar{F}$ is complete.

2.5. Algebras and ideals

A polynormed algebra is a polynormed vector space fitted out with a structure of algebra such that the multiplication is a bounded linear mapping. This means that the boundedness satisfies condition

AB) The product of two bounded sets is bounded.

A convex boundedness on an algebra $A$ which satisfies condition AB) is called an algebra boundedness.

For instance, $\mathcal{C}(\delta)$, $\mathcal{O}(\delta)$, $\mathcal{C}_r(\delta)$, $\mathcal{C}\mathcal{E}_r(\delta)$, $\mathcal{C}(\Delta)$ or $\mathcal{O}(\Delta)$ are polynormed algebras. More precisely, the identity mapping

$${}_N\mathcal{C}(\delta) \cdot {}_P\mathcal{C}(\delta) \to {}_{N+P}\mathcal{C}(\delta)$$

is continuous for all $N, P \in \mathbf{Z}$ and the similar properties are valid for $\mathcal{O}(\delta)$, $\mathcal{C}_r(\delta)$, $\mathcal{C}\mathcal{E}_r(\delta)$.

If $\underline{A}$ is a polynormed algebra, also $\mathcal{C}(\delta; \underline{A})$, $\mathcal{O}(\delta; \underline{A})$, $\mathcal{C}_r(\delta; \underline{A})$, $\mathcal{C}\mathcal{E}_r(\delta; \underline{A})$ are .

Let $\underline{A} = (A, ({}_iA)_{i\in I})$ be a polynormed algebra and $A[X]$ denote the algebra of polynomials with coefficients in A. We consider on $A[X]$ the covering by all vector subspaces

$$ {}_iA + {}_iA \, . \, X + \ldots + {}_iA \, . \, X^N , $$

identified with products ${}_iA^{N+1}$, when i varies in I and N in $\mathbf{N}$. Thus $A[X]$ is a polynormed algebra; a subset B of $A[X]$ is bounded if the degrees and coefficients of elements of B are.

We define similarly $A[X_1,\ldots,X_n]$. For instance, the polynormed algebras $\mathcal{O}(\delta_o)$ and $\mathbf{C}[X_1,\ldots,X_n]$ are isomorphic; this is a consequence of Liouville's Theorem.

A vector space E equipped with a complete convex boundedness (that is a class of complete polynormed vector spaces which are isomorphic above E) is called a b-space. An algebra equipped with a complete algebra boundedness is called a b-algebra. We have already found many examples of b-algebras; note that if A is a b-algebra, also $\mathcal{C}(\delta; A)$ or $A[X_1,\ldots,X_n]$ are.

An ideal I of a commutative b-algebra A, equipped with a complete convex boundedness, is said to be a b-ideal if both the identity mapping $I \to A$ and the multiplication $A \times I \to I$ are bounded linear mappings. It is equivalent to ask that every bounded set in I is bounded in A and that the product of a bounded set in A by a bounded set in I is bounded in I.

Let $a_1,\ldots,a_p$ be elements of a commutative b-algebra A. We equip the ideal

$$ I = \mathrm{idl}(a_1,\ldots,a_p; A) $$

generated by $a_1,\ldots,a_p$ with a structure of b-ideal as following. Assume that the boundedness of A is associated with a covering $({}_iA)_{i\in I}$ by Banach spaces. We consider the covering of I defined by

$$ {}_iI = a_1 \, . \, {}_iA + \ldots + a_p \, . \, {}_iA , $$

where ${}_iI$ is identified with the quotient

$$ {}_iA \times \ldots \times {}_iA / \mathrm{Ker}\, \varphi_i $$

of the product ${}_iA \times \ldots \times {}_iA$ by the kernel of the linear mapping

$$ \varphi_i : (x_1,\ldots,x_p) \to a_1 x_1 + \ldots + a_p x_p . $$

It is easily seen that the identity mapping $I \to A$ and the multiplication $A \times I \to I$ are bounded linear mappings. From the first property, we deduce that I is Hausdorff as A is. Then, each ${}_iI$ is a Banach space and I is complete.

We have to prove that the boundedness of I only depends on the boundedness of A. Note that a set B in I is bounded if it is the image by some $\varphi_i$ of a bounded set of ${}_iA \times \ldots \times {}_iA$. This means that there exist bounded sets $B_1,\ldots,B_p$ in A such that

$$ B \subset a_1 B_1 + \ldots + a_p B_p , $$

(of course, we may choose $B_1 = \ldots = B_p$).

Similarly, when $I_1,\ldots,I_n$ are b-ideals of a b-algebra A, there is a natural way to equip $I_1+\ldots+I_n$ with a structure of b-ideal. A subset B of I is bounded if there exist bounded sets $B_1,\ldots,B_n$ in $I_1,\ldots,I_n$ such that $B \subset B_1+\ldots+B_n$.

Let I be an ideal of a commutative Banach algebra A with unit element. Then $I = A$ as soon as 1 is the limit of a sequence of I, because the set of invertible elements is a neighbourhood of 1. The statement is no longer valid when A is a b-algebra. However

Proposition 4 (L. Waelbroeck).- Let I a b-ideal of a commutative b-algebra A with unit element. Then $I = A$ if 1 is the limit in A of a bounded sequence in I, or, more generally, if there exists a sequence $(x_p)$ such that $x_p = O(k_1^p)$ in I, $1-x_p = O(k_2^p)$ in A, with $k_1 k_2 < 1$.

Proof. We only have to show that 1 belongs to I; if such a property holds, as the multiplication by 1 gives a morphism $A \to I$, we obtain the equality $I = A$ between b-spaces.

Setting $y_p = 1-x_p$, and writing

$$x_{p+1}-x_p = y_p x_{p+1} - x_p y_{p+1},$$

we get $x_{p+1} - x_p = O((k_1 k_2)^p)$ in I. Therefore $\sum_{p\geq 0} (x_{p+1}-x_p)$ converges in I. But

$$\sum_{p\geq 0} (x_{p+1}-x_p) = 1 - x_0$$

in A; hence $1 \in I$.

We also need a more precise result.

Proposition 5.- We consider a b-ideal I of a commutative b-algebra A with unit element, an absolutely convex bounded set B in I, a normed space E of the definition of A and a Banach space F of the definition of I such that $E_B$ and $E \times E_B$ are continuously mapped into F. If 1 is the limit in E of a sequence of B, then 1 belongs to the closure of B in F.

Proof. We may assume that $\|yx\|_F \leqslant \|x\|_E$ for all $x \in E$, $y \in B$. For every $\varepsilon > 0$, we choose a sequence $(x_p)$ in B such that $y_p = 1-x_p$ satisfies $\|y_p\|_E \leqslant \varepsilon 2^{-p-2}$. Then

$$\|x_{p+1}-x_p\|_F = \|y_p x_{p+1} - x_p y_{p+1}\| \leqslant \varepsilon 2^{-p-1},$$

and $\sum_{p\geqslant 0} (x_{p+1}-x_p)$ converges in F; therefore $1 \in F$ and

$$\|1-x_0\|_F = \Big\|\sum_{p\geqslant 0} (x_{p+1}-x_p)\Big\|_F \leqslant \varepsilon .$$

Notes.

(*) A pseudonormed vector space is a vector space equipped with a finite pseudonorm.

Bounded structures have been first studied in a systematical way by L. Waelbroeck (1). The exposition is different here because we put the emphasis on the family of pseudonorms which defines the convergence. Such a point of view is fitted to the examples and problems we shall consider. Actually, the category of polynormed vector spaces is only equivalent to the category of vector spaces equipped with convex boundedness. For bounded structures and their application to functional analysis, the reader is also referred to C. Houzel (1), H. Buchwalter (1), H. Hogbe-Nlend (1) and L. Waelbroeck (4), (5). Although most of the algebras we use are Hausdorff and even complete, we consider pseudonorms when defining polynormed vector spaces, because algebras $\mathcal{O}(\Delta)$ and $\mathcal{O}(K)$ are not necessarily Hausdorff.

The boundedness on $\overline{F}$ when $F$ is a vector subspace has been used by the author in (2). It is an improvement of the previous consideration of the closure, in view of approximation problems. More information about the limiting operations which lead to the closure of a subspace is given by L. Waelbroeck (5).

Proposition 4 is called "Fundamental Lemma" by L. Waelbroeck (1), (3).

# CHAPTER III

## SPECTRAL THEORY OF b-ALGEBRAS

We define the spectrum of one or several elements in a commutative algebra A with unit element. The case of Banach algebras is first discussed and the elementary properties recalled. In the case of b-algebras, the consideration of the algebraic spectrum is not sufficient. We define spectral sets and spectral functions. A subset S of $\mathbf{C}^n$ is said to be spectral for $a_1, \ldots, a_n$ if there exists a bounded set B such that $(a_1 - s_1)B + \ldots + (a_n - s_n)B$ contains 1 for all $(s_1, \ldots, s_n)$ in the complement of S. The concept of a spectral function is a refinement of that of a spectral set. When A is a Banach algebra, a subset S of $\mathbf{C}^n$ (resp. a non negative function $\delta$ on $\mathbf{C}^n$) is spectral for $a_1, \ldots, a_n$ if and only if it is a neighbourhood of the algebraic joint spectrum (resp. is locally bounded from zero on the algebraic joint spectrum). We prove, in the general case, that every spectral function is larger than some spectral function which is a weight function. For every weight function $\delta$, spectral for $a_1, \ldots, a_n$, we construct a bounded linear mapping $f \mapsto f[a]$ from $\mathcal{O}(\delta)$ into A which maps $p(z_1, \ldots, z_n)$ onto $p(a_1, \ldots, a_n)$ for every polynomial p. This is the holomorphic functional calculus. We also introduce a b-ideal I of A, consider spectral functions modulo I, and construct an holomorphic functional calculus which is a mapping from $\mathcal{O}(\delta)$ into A/I. We prove that, when $a_i \equiv b_i$ modulo I, spectral functions modulo I for $a_1, \ldots, a_n$ and $b_1, \ldots, b_n$ are the same and that $f[a_1, \ldots, a_n]$ and $f[b_1, \ldots, b_n]$ are equal in A/I. We also prove that 0 is never spectral for $a_1, \ldots, a_n$ modulo I unless A = I.

## 3.1.- Spectrum of elements in a Banach algebra

We shall only consider commutative algebras A with unit element. This assumption will not be explicitly mentioned. Most of the results remain however valid when A is not commutative, for elements taken in the center of A.

The spectrum of an element a of A is the set of all complex numbers s such that a-s has no inverse. It is denoted by sp(a; A) or sp(a).

More generally, let $a_1,\dots,a_n$ be elements of A. The joint spectrum of $a_1,\dots,a_n$ is the set of all $s = (s_1,\dots,s_n)$ in $\mathbf{C}^n$ such that the ideal

$$\mathrm{idl}\,(a_1 - s_1, \dots, a_n - s_n;\ A),$$

generated by $a_1 - s_1, \dots, a_n - s_n$ in A is different from A. It is denoted by $\mathrm{sp}\,(a_1,\dots,a_n;\ A)$ or $\mathrm{sp}\,(a_1,\dots,a_n)$.

We now consider a Banach algebra, that is an algebra A with a Banach norm such that

$$\| xy \| \leqslant \| x \| \,.\, \| y \|,$$

for every x, y in A. It is well known and easily shown that the set of invertible elements is an open neighbourhood of the origin and that the mapping $x \mapsto x^{-1}$ is continuous and even analytic.

Proposition 1.- Let $a_1,\dots,a_n$ be elements of a Banach algebra A.

a) The spectrum $\mathrm{sp}(a_1,\dots,a_n;\ A)$ is a compact subset of $\mathbf{C}^n$.

b) We can find $\mathcal{C}^\infty$ mappings $u_1,\dots,u_n$ defined on the complement of $\mathrm{sp}(a_1,\dots,a_n)$ and taking their values in A such that $u_i(s) = O(|s|^{-1})$ at infinity for $i=1,\dots,n$ and

$$(a_1 - s_1)\, u_1(s) + \dots + (a_n - s_n)\, u_n(s) = 1,$$

for every $(s_1,\dots,s_n)$ in the complement of $\mathrm{sp}(a_1,\dots,a_n)$.

Proof. It is easily seen that there exists an open neighbourhood $V(\infty)$ of infinity such that $(a_1 - s_1)\bar{s}_1 + \dots + (a_n - s_n)\bar{s}_n$ is invertible for s in $V(\infty)$. Setting

$$w_{i,\infty}(s) = \bar{s}_i\,((a_1 - s_1)\bar{s}_1 + \dots + (a_n - s_n)\bar{s}_n)^{-1},$$

we have

$$(a_1 - s_1) w_{1,\infty}(s) + \dots + (a_n - s_n) w_{n,\infty}(s) = 1.$$

Thus $\mathrm{sp}(a_1,\dots,a_n)$ is contained in the complement of $V(\infty)$ and therefore bounded. Now fix $(t_1,\dots,t_n)$ in the complement of $\mathrm{sp}(a_1,\dots,a_n)$ and choose elements of A such that

$$(a_1 - t_1)\, v_{1,t} + \dots + (a_n - t_n)\, v_{n,t} = 1\ .$$

As the set of invertible elements is open, there exists an open neighbourhood $V(t)$ of $t$ such that $(a_1-s_1)v_{1,t}+\ldots+(a_n-s_n)v_{n,t}$ is invertible for $s$ in $V(t)$. Setting now

$$w_{i,t}(s) = v_{i,t}((a_1-s_1)v_{1,t}+\ldots+(a_n-s_n)v_{n,t})^{-1},$$

we have

$$(a_1-s_1)\,w_{1,t}(s)+\ldots+(a_n-s_n)\,w_{n,t}(s) = 1.$$

Then $V(t)$ is contained in the complement of $sp(a_1,\ldots,a_n)$. Therefore property a) is proved.

Choose now a $\mathcal{C}^\infty$ partition of unit $\varphi_\infty$, $(\varphi_t)$ subordinated to the covering $V(\infty)$, $(V(t))$ of the complement of $sp(a_1,\ldots,a_n)$ such that $\varphi_\infty(s)=1$ on a neighbourhood of infinity. Obviously, each

$$u_i = \varphi_\infty w_{i,\infty} + \sum_t \varphi_t\, w_{i,t}$$

is $\mathcal{C}^\infty$ and

$$(a_1-s_1)\,u_1(s)+\ldots+(a_n-s_n)\,u_n(s) = 1.$$

Moreover, in a neighbourhood of infinity, we have

$$u_i(s) = w_{i,\infty}(s) = O(|s|^{-1}),$$

and the proof of b) is complete.

A well-known property is the fact that $sp(a_1,\ldots,a_n)$ is never empty. When $n=1$, this follows from Liouville's Theorem as the resolvent function

$$s \mapsto (a-s)^{-1}$$

is analytic on the complement of $sp(a)$ in the Riemann sphere. We shall give a proof of the property in a more general setting at the end of the Chapter. This can also be deduced from the consideration of the set $M$ of all maximal ideals of $A$.

We identify $M$ with the set of multiplicative linear forms which do not identically vanish. The kernel of a multiplicative linear form $\chi \neq 0$ is a maximal ideal. Conversely, if $m$ is a maximal ideal, $m$ does not intersect the set of invertible elements and is therefore closed. The quotient space $A/m$ is a Banach algebra and a field. Then $A/m = \mathbf{C}$ because every element which does not lie in $\mathbf{C}$ has an empty spectrum, and $m$ is the kernel of the multiplicative linear form

$$A \to A/m = \mathbf{C}.$$

As usual $M$ is equipped with the weakest topology such that the mapping

$$\hat{a} : \chi \mapsto \chi(a)$$

is continuous for every $a \in A$. This identifies $M$ with a closed subset of the product space

$$\prod_{a \in A} \{|z| \leq \|a\|\}$$

and therefore M is a compact space.

Proposition 2.- *Let* $a_1, \ldots, a_n$ *be elements of a Banach algebra* A. *Then* $sp(a_1, \ldots, a_n)$ *is the set of elements*

$$(\chi(a_1), \ldots, \chi(a_n)),$$

*when* $\chi$ *ranges over* M.

It is obvious that every $(\chi(a_1), \ldots, \chi(a_n))$ belongs to M. Conversely, if $idl(a_1 - s_1, \ldots, a_n - s_n; A)$ is different from A, it is contained in a maximal ideal m. Let $\chi$ be the multiplicative linear form associated to m. We have $\chi(a_i - s_i) = 0$ for $i = 1, \ldots, n$ and then $(s_1, \ldots, s_n) = (\chi(a_1), \ldots, \chi(a_n))$.

As M is not empty, we deduce from Proposition 2 that nor $sp(a_1, \ldots, a_n)$ is.

## 3.2.- Spectral sets

Spectra of elements in b-algebras are not necessarily compact. For instance, as $\mathcal{O}(\delta_o)$ is the algebra of polynomials, we have $sp(z; \mathcal{O}(\delta_o)) = \mathbf{C}$; if D is the unit open disc in the complex plane, the spectrum of z in $\mathcal{O}(\delta_D)$ is the unit disc itself.

We first consider the spectrum of one element a of a b-algebra A. We cannot prove nice properties for the resolvent function $s \mapsto (a-s)^{-1}$ on the complement of $sp(a)$ and have to set a new definition.

Definition 1.- *A subset* S *of* $\mathbf{C}$ *is said to be spectral for* a *in* A, *if* $(a-s)^{-1}$ *exists and is bounded when* s *ranges over the complement of* S.

The set of all spectral subsets S is denoted by $\sigma(a; A)$ or $\sigma(a)$.

Proposition 3.- *The interior of every spectral set for* a *is spectral for* a; *the resolvent* $s \mapsto (a-s)^{-1}$ *is holomorphic in the exterior of every spectral set for* a.

*Proof*. Let $S \in \sigma(a)$. We can find a bounded set B in A such that $(a-s)^{-1}$ exists and belongs to B for every s off S. If s is on the boundary of S, this is the limit of a sequence $(s_p)$ of the complement of S. We have

$$(a-s_p)^{-1} - (a-s_{p+q})^{-1} = (s_p - s_{p+q})(a-s_p)^{-1}(a-s_{p+q})^{-1},$$

and if E is a Banach space of the definition of A such that B and B . B are bounded in E, obviously $(a-s_p)^{-1}$ is a Cauchy sequence in E. Therefore $(a-s_p)^{-1}$ has a limit x in E such that $(a-s)x = 1$ in A, and $a-s$ is invertible. Moreover, when s ranges over the boundary of $S$, it is clear that $(a-s)^{-1}$ remains in a bounded subset of $E$. Thus the interior $\mathring{S}$ of $S$ also belongs to $\sigma(a)$.

Let us consider now an interior point $s$ of the complement of $S$. For $s$ close enough to $s_o$, $(a-s)^{-1}$ exists and belongs to $B$. Using

$$(a-s)^{-1} - (a-s_o)^{-1} = (s-s_o)(a-s)^{-1}(a-s_o)^{-1},$$

we see that $s \mapsto (a-s)^{-1}$ is continuous from $\complement \overline{S}$ into $E$. As $A$ is a b-algebra, $(s,t) \mapsto (a-s)^{-1}(a-t)^{-1}$ is also continuous from $\complement \overline{S} \times \complement \overline{S}$ into some Banach space $F$ of the definition of $A$ such that the identity mapping is continuous from $E$ into $F$. Then the resolvent function $s \mapsto (a-s)^{-1}$ is a complex differentiable mapping taking its values in $F$, and its derivative at $s_o$ is equal to $(a-s_o)^2$; it is even continuously differentiable as a mapping of $\complement \overline{S}$ into $F$.

It follows from Liouville's Theorem and the second part of Proposition 3 that $\emptyset$ is never spectral for $a$. Hence $\sigma(a)$ is a true filter in the complex plane, the intersection of which is $sp(a)$. Moreover $\sigma(a)$ has a basis of open sets.

When $A$ is a Banach algebra, $\sigma(a)$ consists of all neighbourhoods of $sp(a)$ : the resolvent function is bounded on the complement of every neighbourhood of $sp(a)$ and conversely, for every $S \in \sigma(a)$, the interior of $S$ belongs to $\sigma(a)$ and therefore contains $sp(a)$. This is not valid for b-algebras; in that case, $\sigma(a)$ gives much more information than $sp(a)$.

We now define the joint spectrum of elements $a_1, \ldots, a_n$ of a b-algebra $A$.

Definition 2.- A subset $S$ of $\mathbf{C}^n$ is said to be spectral for $a_1, \ldots, a_n$ if one can associate to every $s = (s_1, \ldots, s_n)$ in the complement of $S$, elements $u_1(s), \ldots, u_n(s)$ bounded independently of $s$ in $A$ such that

$$(a_1 - s_1)\, u_1(s) + \ldots + (a_n - s_n)\, u_n(s) = 1.$$

The set of all spectral subsets for $a_1, \ldots, a_n$ is denoted by $\sigma(a_1, \ldots, a_n; A)$ or $\sigma(a_1, \ldots, a_n)$. We shall prove at the end of the Chapter and in a more general setting that $\emptyset$ never belongs to $\sigma(a_1, \ldots, a_n)$. Thus $\sigma(a_1, \ldots, a_n)$ is a true filter in $\mathbf{C}^n$.

We note that $S$ is spectral for $a_1, \ldots, a_n$ if there exists a bounded set $B$ such that $1$ belongs to $(a_1-s_1)B + \ldots + (a_n-s_n)B$ for every $(s_1, \ldots, s_n)$ in the complement of $S$. This condition can be weakened as follows

Proposition 4.- In order that a subset $S$ of $\mathbf{C}^n$ is spectral for $a_1, \ldots, a_n$, it suffices that there exist a bounded set $B$ and a normed space $E$ of the definition of $A$ such that $1$ belongs to the closure in $E$ of $[(a_1-s_1)B + \ldots + (a_n-s_n)B] \cap E$ for every

$(s_1, \ldots, s_n)$ in the complement of $S$.

Proof. First fix $s = (s_1, \ldots, s_n)$ in $\complement S$. Our assumption shows that $1$ is the limit in $A$ of a sequence of $(a_1 - s_1)B + \ldots + (a_n - s_n)B$, that is a bounded sequence of the ideal idl $(a_1 - s_1, \ldots, a_n - s_n; A)$. It follows then from Proposition 4 of Chapter II that $1$ belongs to such an ideal; hence there exist elements $u_1(s), \ldots, u_n(s)$ in $A$ such that

$$(a_1 - s_1)\, u_1(s) + \ldots + (u_n - s_n)\, u_n(s) = 1.$$

The proof will be complete if we show that $u_1(s), \ldots, u_n(s)$ can be chosen in a bounded set independent of $S$. Let $F$ be a Banach space of the definition of $A$ such that $E_B$ and $E \times E_B$ are continuously mapped into $F$. Let

$$B_1 = (a_1 - s_1)B + \ldots + (a_n - s_n)B$$

and let $F_1$ be the Banach space $(a_1 - s_1)F + \ldots + (a_n - s_n)F$ equipped with the norm considered in Section 2. Obviously $E_{B_1}$ and $E \times E_{B_1}$ are continuously mapped into $F_1$. It follows from Proposition 5 of Chapter II that $1$ belongs to the closure of $B_1$ in $F_1$. Then if $C$ is the unit ball of $F$,

$$1 \in B_1 + (a_1 - s_1)C + \ldots + (a_n - s_n)C.$$

In other words

$$1 \in (a_1 - s_1)(B \cup C) + \ldots + (a_n - s_n)(B \cup C)$$

and the statement is proved as $B \cup C$ is independent of $s$.

Proposition 5.- The interior of every spectral set for $a_1, \ldots, a_n$ is spectral for $a_1, \ldots, a_n$.

Proof. Let $S \in \sigma(a_1, \ldots, a_n)$ and choose coefficients $u_1(s)$ satisfying $(a_1 - s_1)\, u_1(s) + \ldots + (a_n - s_n)\, u_n(s) = 1$ and contained in an absolutely convex bounded set $B$. Every point $s = (s_1, \ldots, s_n)$ of $\complement \mathring{S}$ is the limit of a sequence $t_p = (t_{1,p}, \ldots, t_{n,p})$ of $\complement S$. Writing

$$(a_1 - s_1)\, u_1(t_p) + \ldots + (a_n - s_n)\, u_n(t_p) - 1 = (t_{1,p} - s_1)\, u_1(t_p) + \ldots + (t_{n,p} - s_n)\, u_n(t_p),$$

we see that $1$ belongs to the closure of $(a_1 - s_1)B + \ldots + (a_n - s_n)B$ in $E_B$ and Proposition 4 shows that $\mathring{S}$ is spectral for $a_1, \ldots, a_n$.

If $A$ is a Banach algebra, a subset $S$ of $\mathbf{C}^n$ is spectral for $a_1, \ldots, a_n$ if and only if it is a neighbourhood of $\operatorname{sp}(a_1, \ldots, a_n)$ : if $S$ belongs to $\sigma(a_1, \ldots, a_n)$, also $\mathring{S}$ belongs to $\sigma(a_1, \ldots, a_n)$ and contains $\operatorname{sp}(a_1, \ldots, a_n)$; conversely if $S$ is a neighbourhood of $\operatorname{sp}(a_1, \ldots, a_n)$, Proposition 1 shows that $S$ is spectral for $a_1, \ldots, a_n$.

3.3.- Spectral functions

In the study of algebras of entire functions for instance, the consideration of spectral sets gives no information on the algebra. The joint spectrum of the coordinate functions is always $\mathbf{C}^n$. We shall therefore need the following generalization of the spectrum.

Definition 3.- Let $a_1, \ldots, a_n$ be elements of a b-algebra A. A non negative function $\delta$ on $\mathbf{C}^n$ is said to be spectral for a if elements $u_0(s), u_1(s), \ldots, u_n(s)$ of A can be associated to every $s = (s_1, \ldots, s_n)$ in $\mathbf{C}^n$, so that

$$(3.3.1) \qquad (a_1 - s_1)u_1(s) + \ldots + (a_n - s_n)u_n(s) + \delta(s)\, u_0(s) = 1,$$

and $u_1(s), \ldots, u_n(s)$ are bounded in A independently of s.

The set of all spectral functions for a is denoted by $\Delta(a_1, \ldots, a_n; A)$ or $\Delta(a_1, \ldots, a_n)$.

Obviously, $1 \in \Delta(a_1, \ldots, a_n)$: choose $u_0(s) = 1$ and $u_1(s) = \ldots = u_n(s) = 0$. A spectral function $\delta$ gives some information at points s in $\mathbf{C}^n$ such that $\delta(s) = 0$, or such that $\delta$ decreases more or less rapidly near s.

Proposition 6.- Let $a_1, \ldots, a_n$ be elements of b-algebra A,

a) $\delta_0 \in \Delta(a_1, \ldots, a_n)$

b) if $\delta, \delta'$ belong to $\Delta(a_1, \ldots, a_n)$, also $\mathrm{Min}(\delta, \delta')$ belongs to $\Delta(a_1, \ldots, a_n)$

c) if $\delta$ belongs to $\Delta(a_1, \ldots, a_n)$ and $\delta' \geq \varepsilon \delta^N$ for some positive integer N and some positive number $\varepsilon$, then $\delta'$ belongs to $\Delta(a_1, \ldots, a_n)$.

Proof. a) Set

$$u_i(s) = -\bar{s}_i\, \delta_0^2(s) \qquad \text{for } i = 1, \ldots, n,$$

and

$$u_0(s) = (a_1 \bar{s}_1 + \ldots + a_n \bar{s}_n + 1)\, \delta_0(s).$$

Obviously $u_0(s), u_1(s), \ldots, u_n(s)$ are bounded independently of s and

$$(a_1 - s_1)u_1(s) + \ldots + (a_n - s_n)\, u_n(s) + \delta_0(s)\, u_0(s) = 1.$$

b) Let $u_0, u_1, \ldots, u_n$ (resp. $u_0', u_1', \ldots, u_n'$) be associated to $\delta$ (resp. $\delta'$). We set $u_i''(s) = u_i(s)$ for $i = 0, 1, \ldots, n$ if $\delta(s) \leq \delta'(s)$ and $u_i''(s) = u_i'(s)$ for $i = 0, 1, \ldots, n$ if $\delta'(s) < \delta(s)$. Coefficients $u_i''(s)$ are bounded in A independently of s and satisfy

$$(a_1 - s_1)\, u_1''(s) + \ldots + (a_n - s_n)\, u_n''(s) + \mathrm{Min}(\delta(s), \delta'(s))\, u_0''(s) = 1.$$

c) We first prove that if $\delta \in \Delta(a_1, \ldots, a_n)$ and $\delta' \geq \varepsilon\delta$ for some positive number

$\varepsilon$ , then $\delta' \in \Delta(a_1, \ldots, a_n)$. If $u_o, u_1, \ldots, u_n$ are associated to $\delta$, we keep $u_1, \ldots, u_n$ and take $u_o'(s) = 0$ if $\delta'(s) = 0$ and $u_o'(s) = \frac{\delta(s)}{\delta'(s)} u_o(s)$ if $\delta'(s) > 0$. Coefficients $u_o', u_1, \ldots, u_n$ easily satisfy the required conditions for $\delta'$.

We only have to show that $\delta^N \in \Delta(a_1, \ldots, a_n)$ if $\delta \in \Delta(a_1, \ldots, a_n)$ and N is a positive integer. Thanks to the properties already proved, we may assume that $\delta$ is bounded. Taking

$$(a_1 - s_1)\, u_1(s) + \ldots + (a_n - s_n)\, u_n(s) + \delta(s)\, u_o(s) = 1$$

at the $N^{th}$ power, we get

$$U(s) + \delta^N(s)\, u_o^N(s) = 1,$$

where $U(s)$ is obviously bounded independently of $s$ in $\mathrm{idl}(a_1 - s_1, \ldots, a_n - s_n; A)$.

It follows from Proposition 6 that every non negative function is spectral as soon as it is equivalent to a spectral function.

A subset $\Delta_1$ of $\Delta(a_1, \ldots, a_n)$ is said to be a <u>basis</u> of $\Delta(a_1, \ldots, a_n)$ if every function $\delta$ in $\Delta(a_1, \ldots, a_n)$ is larger than some function in $\Delta_1$.

Proposition 7.- <u>A basis of</u> $\Delta(a_1, \ldots, a_n)$ <u>consists of all functions</u> $\varphi_B$, <u>where</u> B <u>is an absolutely convex bounded set in</u> A <u>and</u> $\varphi_B(s)$ <u>the distance in</u> $E_B$ <u>from</u> 1 <u>to</u>

$$(a_1 - s_1)\, B + \ldots + (a_n - s_n)\, B,$$

<u>and such functions are Lipschitz over</u> $\mathbf{C}^n$.

<u>Proof</u>. It is easily seen that if $\delta$ is spectral for $a_1, \ldots, a_n$, there exists some absolutely convex bounded set B such that $\delta \geqslant \varphi_B$; we only have to choose B large enough so that it contains $u_o(s), u_1(s), \ldots, u_n(s)$.

Further, each $\varphi_B$ is spectral for $a_1, \ldots, a_n$. Let S denote the set where $\varphi_B$ does not vanish. For every point $s \notin S$, the closure of $(a_1 - s_1) B + \ldots + (a_n - s_n) B$ in $E_B$ contains 1. It follows then from Proposition 4 that S is spectral for $a_1, \ldots, a_n$. When $\varphi_B(s) = 0$, we thus can find coefficients $u_o(s), \ldots, u_n(s)$ which are bounded independently of s and satisfy

$$(a_1 - s_1)\, u_1(s) + \ldots + (a_n - s_n)\, u_n(s) = 1.$$

When $\varphi_B(s) > 0$, it follows immediately from the definition of $\varphi_B(s)$ that there exists some $u_o(s) \in B$ such that $1 + 2\varphi_B(s)\, u_o(s)$ belongs to $(a_1 - s_1) B + \ldots + (a_n - s_n) B$. Thus we can find $u_1(s), \ldots, u_n(s)$ in B such that

$$(a_1 - s_1) u_1(s) + \ldots + (a_n - s_n) u_n(s) + \varphi_B(s) \,.\, 2 u_o(s) = 1.$$

Now let $s, s'$ be different points in $\mathbf{C}^n$ and $\varepsilon$ a positive number. There exist elements $u_o(s), u_1(s), \ldots, u_n(s)$ of B such that

$$(a_1 - s_1) u_1(s) + \ldots + (a_n - s_n) u_n(s) + (\varphi_B(s) + \varepsilon) u_o(s) = 1.$$

Then, we also have

$$(a_1 - s_1') u_1(s) + \ldots + (a_n - s_n') u_n(s) + (\varphi_B(s) + \varepsilon) u_o(s) + (s_1' - s_1) u_1(s) + \ldots + (s_n' - s_n) u_n(s) = 1,$$

and $(a_1 - s_1') u_1(s) + \ldots + (a_n - s_n') u_n(s) - 1$ belongs to the set

$$(\varphi_B(s) + \varepsilon + |s_1' - s_1| + \ldots + |s_n' - s_n|) B.$$

Hence

$$\varphi_B(s') \leqslant \varphi_B(s) + \varepsilon + |s_1' - s_1| + \ldots + |s_n' - s_n|$$

and, as $\varepsilon$ is arbitrary

$$\varphi_B(s') \leqslant \varphi_B(s) + |s_1' - s_1| + \ldots + |s_n' - s_n|,$$

which shows that $\varphi_B$ is Lipschitz over $\mathbf{C}^n$.

Corollary 1.- <u>If</u> $\delta$ <u>is spectral for</u> $a_1, \ldots, a_n$, <u>also</u> $\tilde{\delta}$ <u>is</u>.

There exists some absolutely convex bounded set B such that $\delta \geqslant \varphi_B$ and therefore $\delta \geqslant \frac{1}{n} \varphi_B$. As $\frac{1}{n} \varphi_B$ satisfies condition W2, we have $\tilde{\delta} \geqslant \frac{1}{n} \varphi_B$ and $\tilde{\delta}$ is spectral for $a_1, \ldots, a_n$.

If $\delta$ is spectral for $a_1, \ldots, a_n$, also $\operatorname{Min}(\tilde{\delta}, \delta_o)$ is. Therefore $\Delta(a_1, \ldots, a_n)$ has always a basis of weight functions.

It is obvious that for every spectral function $\delta$, the set $\{\delta > 0\}$ is spectral. Conversely, if S is a spectral set, the characteristic function $\chi_S$ is spectral. Therefore

$$\delta_S = \operatorname{Min}(\tilde{\chi}_S, \delta_o)$$

is also a spectral function.

## 3.4.- <u>The holomorphic functional calculus</u>

We consider a b-algebra A and elements $a_1, \ldots, a_n$ in A. We shall write $(a_1, \ldots, a_n) = a$, $\Delta(a_1, \ldots, a_n) = \Delta(a)$ and $x_1 y_1 + \ldots + x_n y_n = \langle x, y \rangle$, when $x = (x_1, \ldots, x_n)$ and $y = (y_1, \ldots, y_n)$ belong to $A^n$.

Let $\delta$ be a weight function in $\Delta(a)$. Our aim is to construct a bounded linear mapping of $\mathcal{O}(\delta)$ into A which maps $p(z)$ onto $p(a)$ for every polynomial p; this will be the holomorphic functional calculus at a.

We first have to regularize the coefficients which appear in the definition of the spectrum. For every positive integer N, we denote by $S_N(a;\delta;A)$ or $S_N(a;\delta)$ the set of all continuously differentiable systems $u = (u_1,\dots,u_n)$ defined on $\mathbf{C}^n$ and taking their values in A, bounded along with their derivatives of order 1 and such that

$$y = 1 - \langle a-z, u\rangle \tag{3.4.1}$$

belongs to ${}_{-N}\mathcal{C}_1(\delta;A)$.

Lemma 1.- <u>We have</u> $S_{N+1}(a;\delta)\subset S_N(a;\delta)$ <u>and each</u> $S_N(a;\delta)$ <u>is non void</u>.

<u>Proof</u>. The first part of the statement is obvious and we prove the second one for N large enough. It follows from Proposition 4 of Chapter I that there exists a weight function $\delta'$ which satisfies $\frac{1}{2}\delta \leqslant \delta' \leqslant \frac{3}{2}\delta$ and is $\mathcal{E}^\infty$ on the open set $S = \{\delta > 0\}$. It is easily shown that $\delta'^N$ is continuously differentiable on $\mathbf{C}^n$ for $N \geqslant 2$ : for every derivative D of order 1, as $\delta'$ is Lipschitz, clearly $D(\delta'^N) = 0$ on the boundary of S and $D(\delta'^N) = N\delta^{N-1}D\delta'$ tends to zero at the boundary of S. Moreover $\delta' \in {}_{-1}\mathcal{C}_1(\delta)$ and $\delta'^N \in {}_{-N}\mathcal{C}_1(\delta)$. Therefore the statement is an easy consequence of

Lemma 2.- <u>Let</u> $\gamma$ <u>a continuously differentiable weight function in</u> $\Delta(a;A)$; <u>there exist continuously differentiable functions</u> $u_0, u_1,\dots,u_n$ <u>defined on</u> $\mathbf{C}^n$ <u>and taking their values in</u> A, <u>bounded along with their derivatives of order</u> 1 <u>and such that</u>

$$\langle a-z, u\rangle + \gamma u_0 = 1 .$$

<u>Proof</u>. We already know that there exist bounded functions $v_0, v_1,\dots,v_n$ such that

$$\langle a-z, v\rangle + \gamma v_0 = 1.$$

Choose a $\mathcal{E}^\infty$ non negative function $\psi$ on $\mathbf{C}^n$ with support in the unit square

$$|x_1| < 1,\dots,|x_n| < 1,\ |y_1| < 1,\dots,|y_n| < 1$$

and positive on the square

$$|x_1| \leqslant \tfrac{1}{2},\dots,|x_n| \leqslant \tfrac{1}{2},\ |y_1| \leqslant \tfrac{1}{2},\dots,|y_n| \leqslant \tfrac{1}{2}.$$

Let $T = (\mathbf{Z} + i\mathbf{Z})^n$ and

$$\varphi(s) = \psi(s) / \sum_{t\in T} \psi(s+t).$$

Obviously $\varphi$ has the properties already mentioned and $\sum_{t\in T} \varphi(s+t) = 1$. For every positive integer p and $i=0,\dots,n$, we set

$$w_{i,p}(s) = \sum_{t\in T} \varphi(2^p s - t)\, v_i(2^{-p}t).$$

Clearly each $w_{i,p}$ is $\mathcal{E}^\infty$ on $\mathbf{C}^n$ and is bounded independently of $p$, whereas $D w_{i,p} = O(2^p)$ for every derivative $D$ of order 1. Moreover

$$(3.4.2) \qquad \langle a-s,\, w_p(s)\rangle + \gamma(s)\, w_{o,p}(s) + k_p(s) = 1$$

with

$$k_p(s) = \sum_{t\in T} \varphi(2^p s-t)\, \langle 2^{-p}t-s,\, v(2^{-p}t)\rangle + \sum_{t\in T} \varphi(2^p s-t)(\gamma(s)-\gamma(2^{-p}t))\, v_o(2^{-p}t).$$

When $\varphi(2^p s-t)\neq 0$, we have $|s-2^{-p}t| < 2^{-p+1}\sqrt{2n}$; thereby $|\gamma(s)-\gamma(2^{-p}t)| = O(2^{-p})$. Then, as easily shown, $k_p(s) = O(2^{-p})$. Taking now (3.4.2) at the 4th power and arranging terms, we obtain coefficients $W_{o,p}, W_{1,p},\ldots, W_{n,p}$ such that

$$\langle a-z,\, W_p\rangle + \gamma W_{o,p} + k_p^4 = 1.$$

Each $W_{i,p}$ is bounded independently of $p$, whereas $k_p^4 = O(2^{-4p})$. Besides $D W_{i,p} = O(2^p)$ and $D(k_p^4) = O(2^{-2p})$ for every derivative $D$ of order 1. We may apply Proposition 4 of Chapter II to the ideal generated by $a_1-z_1,\ldots, a_n-z_n, \gamma$ in $\mathcal{T}\mathcal{E}_1(\delta_o; A)$: we can find $w_o, w_1,\ldots, w_n$ in $\mathcal{T}\mathcal{E}_1(\delta_o; A)$ such that

$$\langle a-z,\, w\rangle + \gamma w_o = 1.$$

The new coefficients $w_o, w_1,\ldots, w_n$ belong to some ${}_N\mathcal{T}\mathcal{E}_1(\delta_o; A)$; they have polynomial growth at infinity. In order to obtain bounded coefficients, we consider the coefficients

$$U_1 = -\bar z_1\,\delta_o^2,\ldots, U_n = -\bar z_n\,\delta_o^2$$

and

$$Y = (\langle a, \bar z\rangle + 1)\,\delta_o^2$$

of Proposition 6 a). They lie in ${}_{-1}\mathcal{T}\mathcal{E}_1(\delta_o; \mathbf{C})$ and satisfy

$$\langle a-z,\, U\rangle + Y = 1.$$

Therefore

$$1 = \langle a-z,\, U\rangle + Y(\langle a-z,\, w\rangle + \gamma w_o)$$
$$= \langle a-z,\, U + Yw\rangle + \gamma Y w_o\,.$$

Now $U_1 + Yw_1,\ldots, U_n + Yw_n$ and $Yw_o$ lie in ${}_{N-1}\mathcal{T}\mathcal{E}_1(\delta_o; A)$. By decreasing induction on $N$, we easily obtain coefficients $u_o, u_1,\ldots, u_n$ in ${}_{-1}\mathcal{T}\mathcal{E}(\delta_o; A)$; such coefficients are bounded along with their derivatives of order 1.

We introduce now a differential form; writing $dz_1\wedge\ldots\wedge dz_n = \mathbf{dz}$ and $d''u_1\wedge\ldots\wedge d''u_n = \mathbf{d}''u$,

Proposition 8.- Let $u\in S_N(a;\delta)$ and $f\in {}_P\mathcal{O}(\delta)$; then $f\,\mathbf{d}''u\wedge\mathbf{d}\,z$, extended by zero on the complement of the set $S=\{\delta>0\}$, is continuous and integrable over $\mathbf{C}^n$ for $N\geqslant P+2n+2$.

Proof. We first show that $f\mathbf{d}''u$ tends to zero at the boundary of S. Using (3.4.1) we have

$$\mathbf{d}''u = (\langle a-z, u\rangle + y)\,\mathbf{d}''u$$

and

$$\mathbf{d}''u = (a_1-z_1)u_1\, d''u_1 \wedge \ldots \wedge d''u_n + \ldots + d''u_1 \wedge \ldots \wedge (a_n-z_n)u_n\, d''u_n + y\,\mathbf{d}''u .$$

But, differentiating (3.4.1), we get

$$(a_1-z_1)\,d''u_1 + \ldots + (a_n-z_n)\,d''u_n + d''y = 0. \tag{3.4.3}$$

Therefore

$$\mathbf{d}''u = -u_1\, d''y \wedge \ldots \wedge d'u_n - \ldots - d''u_1 \wedge \ldots \wedge (-u_n\, d''y) + y\,\mathbf{d}''u,$$

and, as y and the coefficients of $d''y$ are in ${}_{N-1}\mathcal{C}(\delta; A)$, also the coefficients of $\mathbf{d}''u$ are. When $N \geqslant P+2$, the coefficients of $f\mathbf{d}''u$ are in ${}_{-1}\mathcal{C}(\delta; A)$ and $f\mathbf{d}''u$ tends to zero at the boundary of S. Extended by 0 on the complement of S, clearly $f\mathbf{d}''u$ is continuous. When $N \geqslant P+2n+2$, the coefficients of $f\mathbf{d}''u$ are in ${}_{-(2n+1)}\mathcal{C}(\delta; A)$. As $\delta$ is a weight function, $f\mathbf{d}''u = O(|z|^{-2n-1})$ at infinity and is integrable over $\mathbf{C}^n$

Proposition 9.- Let $f \in {}_P\mathcal{O}(\delta)$. The element

$$f[a] = \left(\frac{-1}{2\pi i}\right)^n n! \int_{\mathbf{C}^n} f\mathbf{d}''u \wedge \mathbf{d}z \qquad (*)$$

of A does not depend on the particular choice of u in $S_N(a;\delta)$ with $N \geqslant P+2n+2$.

We first need

Lemma 3.- Let $u, u' \in S_N(a;\delta)$; there exists a continuous form v of type $(0, n-1)$ with coefficients in ${}_{-N}\mathcal{C}(\delta; A)$ such that $\mathbf{d}''u' - \mathbf{d}''u = d''v$.

Note that v is not assumed to be differentiable: $d''v$ exists in the distributional sense, but the coefficients of $d''v$ are continuous functions because those of $\mathbf{d}''u$, $\mathbf{d}''u'$ are.

Proof. We shall prove the statement when $n = 2$; the general case is quite similar, the calculations being more complicated.

We have

$$\begin{cases} (a_1-z_1)u_1 + (a_2-z_2)u_2 + y = 1 \\ (a_1-z_1)u_1' + (a_2-z_2)u_2' + y' = 1 \end{cases}$$

where $y, y'$ are in some ${}_{-N}\mathcal{C}(\delta; A)$. Therefore

$$u_1'-u_1 = \left[(a_1-z_1)u_1+(a_2-z_2)u_2+y\right]u_1' - \left[(a_1-z_1)u_1'+(a_2-z_2)u_2'+y'\right]u_1$$
$$= (a_2-z_2)(u_2u_1'-u_2'u_1) + yu_1'-y'u_1.$$

Calculating similar for $u_2'-u_2$ and $y'-y$ and setting

$$\begin{cases} \alpha = u_2u_1'-u_2'u_1 \\ \xi_1 = yu_1'-y'u_1 \\ \xi_2 = yu_2'-y'u_2 \end{cases}$$

we get

$$\begin{cases} u_1'-u_1 = (a_2-z_2)\alpha + \xi_1 \\ u_2'-u_2 = -(a_1-z_1)\alpha + \xi_2 \\ y'-y = -(a_1-z_1)\xi_1-(a_2-z_2)\xi_2 \end{cases}$$

As easily seen, $\alpha$ is bounded along with its derivatives whereas $\xi_1$, $\xi_2$ are in $_{-N}\mathcal{T}_1(\delta; A)$. First, consider the case when $\xi_1 = \xi_2 = 0$; then

$$\mathbf{d}''u'-\mathbf{d}''u = (d''u_1+(a_2-z_2)d''\alpha)\wedge(d''u_2-(a_1-z_1)d''\alpha) - d''u_1\wedge d''u_2$$
$$= \left[-(a_1-z_1)d''u_1 - (a_2-z_2)d''u_2\right]\wedge d''\alpha$$
$$= d''y\wedge d''\alpha$$

Thus $\mathbf{d}''u'-\mathbf{d}''u = d''v$, if $v = y\,d''\alpha$. Supposing now that $\alpha=0$, $\xi_2=0$, we have

$$\mathbf{d}''u'-\mathbf{d}''u = (d''u_1+d''\xi_1)\wedge d''u_2-d''u_1\wedge d''u_2$$
$$= d''\xi_1\wedge d''u_2 .$$

Thus $\mathbf{d}''u'-\mathbf{d}''u = d''v$, if $v = \xi_1 d''u_2$. The case when $\alpha=0$, $\xi_1=0$, is similar. To obtain the general case, we only have to decompose the transformation which, starting from $u_1$, leads to $u_2$, through $(u_1+\xi_1, u_2)$ and $(u_1+\xi_1, u_2+\xi_2)$.

Proof of Proposition 9.- Keep the notations of Lemma 3; we also have

$$f\mathbf{d}''u'\wedge \mathbf{d}z - f\mathbf{d}''u\wedge \mathbf{d}z = f d''v\wedge \mathbf{d}z$$
$$= d(fv\wedge \mathbf{d}z).$$

Applying Stokes' formula, as the coefficients of $fv\wedge \mathbf{d}z$ are $O(|z|^{-2n-2})$ at infinity, we get

$$\int_{\mathbf{C}^n} d(fv\wedge \mathbf{d}z) = 0,$$

and the statement is proved.

Note that we only need the fact that $fv \wedge \mathbf{dz} = O(|z|^{-2n})$ at infinity to prove that the integral of $fv \wedge \mathbf{d}z$ on increasing spheres with center at 0 tends to zero. Thus, if

$$\lim_{R\to\infty} \int_{|z|\leq R} f\,\mathbf{d}''u \wedge \mathbf{d}z$$

exists, it is independent of N such that $N \geq P+2n$.

As $S_{N'}(a;\delta) \subset S_N(a;\delta)$ for $N' \geq N$, the element $f[a]$ does not depend on the integer N such that $N \geq P+2n+2$. Moreover $f \mapsto f[a]$ is a bounded linear mapping of $\mathcal{O}(\delta)$ into A. If $\delta, \delta'$ are weight functions such that $\delta' \geq \varepsilon\delta^Q$ for some positive integer Q and some $\varepsilon > 0$, we have $S_{QN}(a;\delta) \subset S_N(a;\delta')$. If f is in $\mathcal{O}(\delta')$, the element $f[a]$, calculated with $\delta$ or $\delta'$, is therefore the same.

Proposition 10.- <u>For every polynomial</u> p, <u>we have</u> $p[a] = p(a)$.

<u>Proof</u>. We first admit $1[a] = 1$. If p is a polynomial, we can find polynomials $Q_1,\dots,Q_n$ with coefficients in A such that

$$p(a) - p(z) = (a_1 - z_1)Q_1(z) + \dots + (a_n - z_n)Q_n(z).$$

Consider some u in $S_N(a;\delta)$, with N large enough. We have

$$(a_i - z_i)\mathbf{d}''u = d''u_1 \wedge \dots \wedge d''u_{i-1} \wedge ((a_i - z_i)d''u_i) \wedge d''u_{i+1} \wedge \dots \wedge d''u_n,$$

and using (3.4.3), we get

$$(a_i - z_i)\mathbf{d}''u = (-1)^i d''v$$

with

$$v = y\,d''u_1 \wedge \dots \wedge d''u_{i-1} \wedge d''u_{i+1} \wedge \dots \wedge d''u_n.$$

Using $1[a] = 1$, we obtain

$$p(a) - p[a] = \left(\frac{-1}{2\pi i}\right)^n n! \int_{\mathbf{C}^n} (p(a) - p(z))\,\mathbf{d}''u \wedge \mathbf{d}z.$$

Then the result follows from

$$\int_{\mathbf{C}^n} (a_i - z_i)\,Q_i(z)\,\mathbf{d}''u \wedge \mathbf{d}z = (-1)^n \int_{\mathbf{C}^n} d(vQ_i\,\mathbf{d}z)$$

and Stokes' formula.

We may calculate $1[a]$ with $\delta_o$. Choosing for $i = 1,\dots,n$, some $u_i$ in $S_N(a_i;\delta_o)$ and setting

$$y_i = 1 - (a_i - z_i)u_i,$$

we have

$$\begin{aligned} 1 &= (a_1 - z_1)u_1 + y_1 \\ &= (a_1 - z_1)u_1 + y_1((a_2 - z_2)u_2 + y_2), \end{aligned}$$

and by an easy induction

$$1 = \langle a-z, v\rangle + y_1 \dots y_n,$$

with $v_1 = u_1$, $v_2 = y_1 u_2, \dots, v_n = y_1 \dots y_{n-1} u_n$. Clearly $v_1, \dots, v_n$ belong to $S_N(a;\delta_o)$. Moreover

$$\mathbf{d}''v = y_1^{n-1} d''u_1 \wedge y_2^{n-2} d''u_2 \wedge \dots \wedge d''u_n .$$

Thus the problem is reduced to the calculation of $\int_{\mathbf{C}} y^k d''u \wedge dz$, where $k$ is a non negative integer and $u$ in $S_N(a;\delta_o)$ with $N$ large enough. For $k = 0$, we may choose $u$ in $S_2(a;\delta_o)$ if we can show that $\int_{|z|\leqslant R} d''u \wedge dz$ has a limit when $R$ tends to infinity. Taking $u = -\bar{z}\,\delta_o^2$, we have

$$\lim_{R\to+\infty} \int_{|z|\leqslant R} d''u \wedge dz = \lim_{R\to+\infty} \int_{|z|\leqslant R} \frac{-d\bar{z}\wedge dz}{(1+|z|^2)^2}$$
$$= -2\pi i.$$

For $k>0$, we use

$$\int_{\mathbf{C}} y^k d''u \wedge dz = \frac{1}{k+1} \int_{\mathbf{C}} d''u \wedge dz.$$

This is a consequence of Stokes' formula and the following statement:

Lemma 4.- Let $u \in S_N(a;\delta)$ and $k$ be a non negative integer. If

$$\omega = \sum_{i=1}^{n} (-1)^{i+1} u_i d''u_1 \wedge \dots \wedge d''u_{i-1} \wedge d''u_{i+1} \wedge \dots \wedge d''u_n,$$

we have

$$(k+1)\, y^k \mathbf{d}''u = (n+k+1)\, y^{k+1} \mathbf{d}''u - d''(y^{k+1}\omega).$$

Proof. Obviously $d''\omega = n\,\mathbf{d}''u$. Write

$$\mathbf{d}''u = ((a_1-z_1)u_1 + \dots + (a_n-z_n)u_n + y)\, d''u_1 \wedge \dots \wedge d''u_n.$$

As

$$(a_i - z_i)\, d''u_i = -d''y - \sum_{j\neq i} (a_j - z_j)\, d''u_j,$$

we have

$$\mathbf{d}''u = \sum_{i=1}^{n} (-1)^i u_i d''y \wedge d''u_1 \wedge \dots \wedge \widehat{d''u_i} \wedge \dots \wedge d''u_n + y\mathbf{d}''u$$
$$= -d''y \wedge \omega + y\,\mathbf{d}''u .$$

Further

$$(k+1)\, y^k \mathbf{d}''u = (n+k+1)\, y^{k+1}\mathbf{d}''u - (k+1)\, y^k d''y \wedge \omega - n y^{k+1}\mathbf{d}''u,$$

and the statement follows as $d''(y^{k+1}\omega) = (k+1)\, y^k d''y \wedge d''\omega + n y^{k+1}\, \mathbf{d}''u$.

Proposition 11.- Let $\delta$ be a weight function in $\Delta(a; A)$ and $f \in \mathcal{O}(\delta)$. If $\chi$ is a bounded multiplicative linear form on A, setting $\chi(a) = (\chi(a_1), \dots, \chi(a_n))$, we have $\delta(\chi(a)) > 0$ and

$$\chi(f[a]) = f(\chi(a)).$$

Proof. Consider u in $S_N(a; \delta; A)$ with N large enough. Then $(\chi(u_1), \dots, \chi(u_n))$ belongs to $S_N(\chi(a); \delta; \mathbf{C})$ and $\delta$ is spectral for $\chi(a)$ in $\mathbf{C}$. Therefore $\delta(\chi(a)) > 0$ and $\chi(f[a])$ is equal to

$$f[\chi(a)] = \left(\frac{-1}{2\pi i}\right)^n n! \int_{\mathbf{C}^n} d''(\chi(u)) \wedge dz.$$

But every neighbourhood of $\chi(a)$ in $\mathbf{C}^n$ is spectral for $\chi(a)$. Choose a polydisc D with center at $\chi(a)$ so that $\bar{D}$ is compact in $\{\delta > 0\}$. Clearly f is the uniform limit of a sequence $(p_n)$ of polynomials on $\bar{D}$.

The holomorphic functional calculus at $\chi(a)$ being a bounded linear mapping from $\mathcal{O}(\delta_D)$ into $\mathbf{C}$, we have

$$f[\chi(a)] = \lim_{n \to \infty} p_n[\chi(a)].$$

Using Proposition 10, we get $p_n[\chi(a)] = p_n(\chi(a))$ and $f[\chi(a)] = f(\chi(a))$.

## 3.5.- Spectral theory modulo a b-ideal

We shall now examine what happens when a b-ideal I of the b-algebra of A is also considered.

Definition 4.- A non negative function $\delta$ on $\mathbf{C}^n$ is said to be spectral for a modulo I, if we can find bounded mappings $u_o, u_1, \dots, u_n$ of $\mathbf{C}^n$ into A and a bounded mapping v of $\mathbf{C}^n$ into I such that

$$\langle a - z, u \rangle + v + \delta u_o = 1.$$

The set of all spectral functions modulo I for a is denoted by $\Delta(a; A/I)$. The properties of $\Delta(a; A)$ proved in Proposition 6 are extended without modifications to $\Delta(a; A/I)$. There also exists a basis of $\Delta(a; A/I)$ composed of Lipschitz functions $\varphi_{BB'}$ where B is a bounded absolutely convex set in A and B' a bounded set in I, defined as follows: $\varphi_{B,B'}(s)$ is the distance in $E_B$ from 1 to

$$(a_1 - s_1)B + \dots + (a_n - s_n)B + B'.$$

Hence $\Delta(a; A/I)$ has a basis of weight functions.

When $\delta$ is a weight function in $\Delta(a; A/I)$ and $f \in \mathcal{O}(\delta)$, we define $f[a]$ in A/I.

The ideas are similar to those of Section 3.4. We denote by $S_N(a;\delta;A/I)$ the set of all functions $(u_1,\dots,u_n, v)$, where $u_1,\dots,u_n$ (resp. v) are continuously differentiable functions on $\mathbf{C}^n$ taking their values in A (resp. I), bounded along with their derivatives of order 1, and such that

$$y = 1 - \langle a-s, u\rangle - v$$

belongs to ${}_{-N}\mathcal{C}_1(\delta; A)$. A straight forward extension of Lemmas 1 and 2 shows that $S_N(a;\delta;A/I)$ is not void. However, if u, v are in $S_N(a;\delta;A/I)$ it is no longer possible to prove that $f\,\mathbf{d}''u \wedge \mathbf{dz}$ can be extended over $\mathbf{C}^n$ so that it is continuous and integrable. But it is true for

$$fy\,\mathbf{d}''u \wedge \mathbf{dz}$$

when $N \geqslant P+2n+1$, and we set

$$f[a] = \frac{(-1)^n}{(2\pi i)^n}(n+1)!\int_{\mathbf{C}^n} fy\,\mathbf{d}''u \wedge \mathbf{dz}. \tag{3.5.1}$$

When $I=0$, because of Lemma 4, this definition is consistent with that of Section 3.4.

In the general case, the right hand side of equality (3.5.1) is independent modulo I of (u, v) in $S_N(a;\delta;A/I)$ with $N \geqslant P+2n+1$. Let us consider the case $n=2$ and keep the notations of Lemma 4 with (u, v) and (u', v') instead of u and u'. We define $\alpha, \xi_1, \xi_2$ as previously and

$$\begin{cases} \zeta_1 = vu_1' - v'u_1 \\ \zeta_2 = vu_2' - v'u_2 \\ \eta = yv' - vy' \end{cases}$$

Then

$$\begin{cases} u_1' - u_1 = (a_2 - z_2)\alpha + \xi_1 + \zeta_1 \\ u_2' - u_2 = -(a_1 - z_1)\alpha + \xi_2 + \zeta_2 \\ v' - v = -(a_1 - z_1)\zeta_1 - (a_2 - z_2)\zeta_2 + \eta \\ y' - y = -(a_1 - z_1)\xi_1 - (a_2 - z_2)\xi_2 - \eta \end{cases}.$$

Only the case where, for instance, $\alpha = 0$, $\xi_2 = 0$, $\zeta_1 = \zeta_2 = 0$, $\eta = 0$ requires some explanation. We have

$$y'\,\mathbf{d}''u' - y\,\mathbf{d}''u = [(y-(a_1-z_1)\xi_1)\,d''(u_1+\xi_1) - y\,d''u_1] \wedge d''u_2$$
$$= [-(a_1-z_1)\xi_1\,d''u_1 + y\,d''\xi_1 - (a_1-z_1)\xi_1\,d''\xi_1] \wedge d''u_2 .$$

As

$$-(a_1-z_1)\,d''u_1 \wedge d''u_2 = d''y \wedge d''u_2 + d''v \wedge d''u_2,$$

we get

$$y'\,\mathbf{d}''u' - y\,\mathbf{d}''u = d''\left[(\xi_1 y - \tfrac{1}{2}(a_1-z_1)\xi_1^2)\,d''u_2\right] + \xi_1\,d''v \wedge d''u_2 .$$

But $\xi_1 d''v \wedge d''u_2 \wedge dz$ is integrable over $\mathbf{C}^2$ in I, whereas, by Stokes'formula,

$$\int_{|z|=R} (\xi_1 y - \tfrac{1}{2}(a_1 - z_1)\xi_1^2)\, d''u_2 \wedge dz$$

tends to zero in A when R tends to infinity. Therefore

$$\int_{\mathbf{C}^2} y' \, d''u' \wedge dz - \int_{\mathbf{C}^2} y \, d''u \wedge dz$$

belongs to I.

We are able now to give a generalization of the fact that the spectrum is never void.

Proposition 12.- If 0 is spectral for $a_1, \ldots, a_n$ modulo I, then I = A.

Proof.- If $0 \in \Delta(a; A/I)$, we may choose $y = 0$. Then (3.5.1) yields $1[a] = 0$ in A/I. As $1[a] = 1$, this implies $1 \equiv 0$ modulo I, that is I = A.

Consider now some $b = (b_1, \ldots, b_n)$ such that $b_i \equiv a_i$ modulo I for i=1, ..., n. If $\delta$ is spectral for a modulo I, there exist bounded mappings $u_o, u_1, \ldots, u_n$ (resp. v) of $\mathbf{C}^n$ into A (resp. I) such that

$$\langle a-z, u\rangle + v + \delta u_o = 1.$$

This can be written

$$\langle b-z, u\rangle + v + \langle a-b, u\rangle + \delta u_o = 1.$$

As $a_i - b_i$ belongs to I for each i, obviously $\langle a-b, u\rangle$ is bounded in I and $\delta$ is also spectral for b modulo I. Moreover $S_N(a; \delta; A/I) = S_N(b; \delta; A/I)$ and $f[a] \equiv f[b]$ mod. I.

We have constructed a linear mapping of $\mathcal{O}(\delta)$ into A/I. There is no natural boundedness on A/I; however the following property holds: if f remains in a bounded subset of $\mathcal{O}(\delta)$ and u, v are chosen in $S_N(a; \delta; A/I)$ for N large enough, the element $f_u[a]$ given by (3.5.1) is bounded in A and $f_u[a] - f_v[a]$ is bounded in I (where $f_v[a]$ is similarly defined).

## Notes

Spectral theory of Banach algebras is due to I.M. Gelfand (1) in the one dimensional case. The construction of the holomorphic functional calculus in the n-dimensional case is due to G.E. Shilov (1), R. Arens and A.P. Calderon (1) and L. Waelbroeck (5). The definition of spectral sets and functions and the construction of the holomorphic functional calculus of b-algebras are given by L. Waelbroeck (1) (2). An abridged version can be found in J.-P. Ferrier (2). The exposition adopted here is slightly different. Our aim has been to give a simple construction of f[a] when there is no ideal. It can be shown that $f \to f[a]$ is an homomorphism, but the proof of L. Waelbroeck ([1]) by means of tensor products is rather complicated. We have omitted such a property here because we shall not use it.

(*) $\mathbf{C}^n$ has been oriented so that $(\frac{1}{2\pi i})^n \, \mathbf{d}\bar{z} \wedge \mathbf{d}z$ is a positive measure.

## CHAPTER IV

# SPECTRAL THEOREMS AND HOLOMORPHIC CONVEXITY

At the beginning of the Chapter a few definitions and elementary properties in connection with plurisubharmonic functions and pseudoconvex domains are briefly recalled. We only insist on the equivalent conditions which lead to the definition of pseudoconvexity. The $L^2$-estimates for the d'' operator are used to study spectral sets for $z_1, \ldots, z_n$ in $\mathcal{O}(\delta)$, when $\delta$ is a weight function. We show that the set $\Omega$ where $\delta$ does not vanish is spectral if and only if it is pseudoconvex and apply this result to discuss spectral functions for z in $\mathcal{O}(\delta)$. The case when $-\log \delta$ is plurisubharmonic in $\Omega$ is first considered. To handle the general case, we introduce a process of plurisubharmonic regularization of weight functions. As an application of spectral theorems, we obtain the existence on every pseudoconvex domain of an holomorphic function with polynomial growth which cannot be extended; we also give a characterization of pseudoconvexity by means of bounded multiplicative linear forms.

## 4.1.- Preliminaries

Let $\Omega$ be an open set in $\mathbf{C}^n$. A function f defined in $\Omega$ with values in $[-\infty, +\infty[$ is called plurisubharmonic if it is upper semi-continuous and if for every complex line L, the restriction of f to $\Omega \cap L$ is subharmonic. A non negative function f defined in $\Omega$ is called log-plurisubharmonic if $\log f$ is plurisubharmonic in $\Omega$.

If f is analytic in $\Omega$, then $|f|$ is log-plurisubharmonic in $\Omega$. When $f_1$, $f_2$ are log-plurisubharmonic, then $f_1 + f_2$ is also log-plurisubharmonic in $\Omega$, as it is upper semi-continuous and as the property is valid for functions whose logarithm is subharmonic. For instance, if $\delta_o = (1+|z|^2)^{-\frac{1}{2}}$ is defined as in Section 1.2, the function

$-\log \delta_0$ is plurisubharmonic in $\Omega$.

Definition 1.- An open set $\Omega$ in $\mathbf{C}^n$ is called pseudoconvex if the following equivalent conditions are fulfilled:

(i) $s \mapsto -\log d(s, \complement\Omega)$ is plurisubharmonic in $\Omega$.

(ii) There exists a plurisubharmonic function f in $\Omega$ such that $\{f \leqslant c\}$ is relatively compact in $\Omega$ for every real number c.

(iii) For every compact set K in $\Omega$, the hull $\widehat{K}_\Omega$ of K with respect to plurisubharmonic functions in $\Omega$ is compact.

Let K be a compact set in an open subset $\Omega$ of $\mathbf{C}^n$. We denote by $\widehat{K}_\Omega$ the set of all points s in $\Omega$ such that

$$f(s) \leqslant \sup_{\zeta \in K} f(\zeta)$$

for every plurisubharmonic (or log-plurisubharmonic) function f in $\Omega$. It is the hull of K with respect to plurisubharmonic functions in $\Omega$. Obviously $\widehat{K}_\Omega$ is bounded because $-\log \delta_0$ tends to infinity at infinity. It can be easily proved that $\widehat{K}_\Omega$ is closed in $\Omega$. For every $s \notin \widehat{K}_\Omega$, there exists some plurisubharmonic function f defined on $\Omega$ such that $f(s) > c = \sup_{\zeta\in K} f(\zeta)$. A convolution argument by a $\mathcal{C}^\infty$ function which only depends on $|z_1|, \ldots, |z_n|$, shows that f is the limit of a decreasing sequence $(f_n)$ of plurisubharmonic functions which are continuous on a neighbourhood of $K \cup \{s\}$. Hence, for n large enough we get

$$\sup_{\zeta\in K} f_n(\zeta) < f_n(s),$$

and the property remains valid on a neighbourhood of s.

We first assume condition (i); then $-\log \delta_\Omega$ is also plurisubharmonic in $\Omega$ as

$$-\log \delta_\Omega(s) = \operatorname{Max}(-\log d(s, \complement\Omega), -\log \delta_0(s)),$$

and condition (ii) follows as $\{\delta_\Omega \leqslant c\}$ is compact in $\Omega$ for every real number c.

If (ii) holds, for every compact subset K of $\Omega$, setting

$$c = \sup_{s\in K} |f(s)|,$$

obviously $\widehat{K}_\Omega$ is contained in $\{f \leqslant c\}$; therefore $\widehat{K}_\Omega$ is relatively compact in $\Omega$ and we obtain (iii).

We only have to prove now that (iii) implies (i). It suffices to show that, if D is the unit open disc in the complex plane, for every $a, b \in \mathbf{C}^n$ such that

$$a + \overline{D} b \subset \Omega,$$

and every subharmonic function $\varphi$ on a neighbourhood of $a + \overline{D} b$ such that

$$-\log d(a + \zeta b) \leqslant \varphi(\zeta) \tag{4.1.1}$$

for every $\zeta \in \partial D$, the same inequality is valid for every $\zeta \in D$. Writing $\varphi$ as the real part of some holomorphic function h, (4.1.1) becomes

$$(4.1.2) \qquad d(a+\zeta b) \geqslant |e^{-h(\zeta)}|.$$

Setting then

$$F_c(\zeta) = a + \zeta b + e^{-h(\zeta)} c ,$$

with $c \in \mathbf{C}^n$, $\|c\| < 1$, we only have to show that $F_c(\partial D) \subset \Omega$ implies $F_c(D) \subset \Omega$. This is true for $c = 0$, and the set of all $\lambda \in [0, 1]$ such that $F_{\lambda c}(D) \subset \Omega$ for a given c, is open. The statement will be proved if it is also closed. Let K be the compact set of all $F_{\lambda c}(\zeta)$ with $\lambda \in [0, 1]$, $\zeta \in \partial D$; for every plurisubharmonic function f in $\Omega$, if $F_{\lambda c}(D) \subset \Omega$, as $f \circ F_{\lambda c}$ is subharmonic, we have

$$f(F_{\lambda c}(\zeta)) \leqslant \sup_{s \in K} f(s),$$

and $F_{\lambda c}(\zeta) \in \widehat{K}_\Omega$ ; thus $F_{\lambda c}(D) \subset \Omega$ implies $F_{\lambda c}(D) \subset \widehat{K}_\Omega$.

Remark. We have considered in Definition 1 the distance d associated to the hermitian norm in $\mathbf{C}^n$; the definition does not change when d is the distance associated to another norm, as the equivalence between conditions (i) and (ii) or (iii) is still valid.

Proposition 1.- Let $\Omega$ be a pseudoconvex open set in $\mathbf{C}^n$ and f be a plurisubharmonic function in $\Omega$ ; the open subset where f is negative is pseudoconvex.

Proof. Let K be a compact set in $\{f < 0\}$. There exists a negative constant c such that $f \leqslant c$ on K. Therefore, $f \leqslant c$ on the hull $\widehat{K}_\Omega$ of K with respect to plurisubharmonic functions in $\Omega$, which is compact because $\Omega$ is pseudoconvex. This implies that $\widehat{K}_\Omega \subset \{f < 0\}$. As $\widehat{K}_\Omega$ contains the hull of K with respect to plurisubharmonic functions in $\{f < 0\}$, the statement is proved.

## 4.2.- Spectral sets for z in $\mathcal{O}(\delta)$.

We denote by $\delta$ a weight function on $\mathbf{C}^n$ and by $\Omega = \{\delta > 0\}$ the set where $\delta$ does not vanish. We want to study the spectrum $\sigma(z; \mathcal{O}(\delta))$ of z in the b-algebra $\mathcal{O}(\delta)$.

A subset S of $\mathbf{C}^n$ is spectral for z in $\mathcal{O}(\delta)$ if, for every $s \notin S$, we can find holomorphic functions $u_1(s), \ldots, u_n(s)$ on $\Omega$ such that

$$(4.2.1) \qquad (z_1 - s_1)\, u_1(s) + \ldots + (z_n - s_n)\, u_n(s) = 1$$

and

$$(4.2.2) \qquad \delta^N |u_i(s)| \leqslant M, \quad i = 1, \ldots, n ,$$

where N is a positive integer and M a positive constant, both independent of s.

When the value of $u_i(s)$ at $\zeta \in \Omega$ is denoted by $u_i(s;\zeta)$, conditions (4.2.1) and (4.2.2) become

(4.2.3) $$(\zeta_1 - s_1)\, u_1(s;\zeta) + \ldots + (\zeta_n - s_n)\, u_n(s;\zeta) = 1$$

and

(4.2.4) $$\delta^N(\zeta)\, |u_i(s;\zeta)| \leqslant M, \quad i=1,\ldots,n$$

for $s \notin S$, $\zeta \in \Omega$.

As the left hand side of (4.2.3) vanishes for $\zeta = s$, every spectral set $S$ for $z$ contains $\Omega$. We are now asking for conditions ensuring that $\Omega$ is spectral for $z$.

It is first easily seen that, if $\Omega$ is spectral for $z$ in $\mathcal{O}(\delta)$, then $\Omega$ is pseudoconvex. We only have to prove that $\widehat{K}_\Omega$ is compact in $\Omega$ for every compact subset $K$ of $\Omega$. We already know that $\widehat{K}_\Omega$ is bounded and closed in $\Omega$. If it is not compact, we can find a sequence $\zeta_n$ in $\widehat{K}_\Omega$ tending to some boundary point $\zeta_0$ of $\Omega$. There exists a positive number $\varepsilon$ such that $\delta \geqslant \varepsilon$ on $K$ and (4.2.4) yields

$$|u_i(s;\zeta)| \leqslant M' = M\,\varepsilon^{-N}$$

for $\zeta \in K$. Then $|u_i(s;\zeta_p)| \leqslant M'$ and $\langle \zeta_p - s, u(s;\zeta_p)\rangle$ tends to zero when $p$ increases. We obtain thus a contradiction with (4.2.3). We shall prove the converse statement; as $\mathcal{O}(\delta)$ contains $\mathcal{O}(\delta_\Omega)$ and $\sigma(z;\mathcal{O}(\delta))$ contains $\sigma(z;\mathcal{O}(\delta_\Omega))$, it suffices to consider the case when $\delta = \delta_\Omega$.

Theorem 1.- Let $\Omega$ be an open set in $\mathbf{C}^n$. Then $\Omega$ is spectral for $z$ in $\mathcal{O}(\delta_\Omega)$ if and only if it is pseudoconvex.

Proof. The necessary condition has already been seen. Assume now that $\Omega$ is pseudoconvex and set $\delta = \delta_\Omega$. We want to find holomorphic functions $u_1(s): \zeta \mapsto u_1(s;\zeta), \ldots, u_n(s): \zeta \mapsto u_n(s;\zeta)$ on $\Omega$ satisfying (4.2.3) and (4.2.4).

It is not difficult to find differentiable functions $u_1(s), \ldots, u_n(s)$; we may set for instance

(4.2.5) $${}_0h_i(s) = \frac{\bar{z}_i - \bar{s}_i}{|z - s|^2},$$

for $i = 1, \ldots, n$. Obviously

(4.2.6) $$\langle z - s, {}_0h(s)\rangle = 1$$

and as

$$\delta(\zeta) = \delta(\zeta) - \delta(s) \leqslant |\zeta - s|$$

for $s \notin \Omega$, $\zeta \in \Omega$, we have

(4.2.7) $$\delta\, |{}_0h_i(s)| \leqslant 1.$$

This implies that $\Omega$ is always spectral for $z$ in $\mathcal{C}_r(\delta)$.

But the coefficients ${}_{o}h_i(s)$ are not holomorphic. In order to get holomorphic functions, we have to modify them. We shall use the following result, if $|u|^2$ denotes the sum of squares of the absolute values of the coefficients of the differential form $u$:

Lemma 1 (L. Hörmander).- Let $\Omega$ be a pseudoconvex open set in $\mathbf{C}^n$, $\varphi$ a plurisubharmonic function in $\Omega$, and $r$ a non negative integer. For every differential form $v$ of type $(0, r)$ in $\Omega$ which is square integrable with respect to the measure $e^{-\varphi}d\lambda$ and satisfies $d''v = 0$ in the distributional sense, there exists a locally integrable differential form $u$ of type $(0, r+1)$ in $\Omega$ such that $d''u = v$ and

$$\int |u|^2 e^{-\varphi}\delta_o^4 \, d\lambda \leqslant \int |v|^2 e^{-\varphi} d\lambda .$$

If $r$, $t$ are non negative integers, let $L_r^t$ denote the vector space of all differential forms of type $(0, r)$ with coefficients in the exterior product $\Lambda^t \mathbf{C}^n$ which are square integrable with respect to some measure $\delta^N d\lambda$. Such a form $h$ is a skew-symetrical system $(h_I)$, where $I$ ranges over the set of all multiindices $(i_1, \dots, i_t)$ with $1 \leqslant i_1, \dots, i_t \leqslant n$. We set

$$|h|^2 = \sum |h_I|^2,$$

where $I$ ranges over the set of multiindices $(i_1, \dots, i_t)$ such that $1 \leqslant i_1 < \dots < i_t \leqslant n$.

We define, for every $s \notin \Omega$, an operator ${}^sP$ from $L_r^{t+1}$ to $L_r^t$ by

$$({}^sPh)_I = \sum_{i=1}^{n} (z_i - s_i) h_{I,i},$$

and set ${}^sPh = 0$ if $h \in L_r^o$. We consider the double complex

$$\begin{array}{ccc} L_r^{t+1} & \xrightarrow{\ d''\ } & L_{r+1}^{t+1} \\ {}^sP \downarrow & & \downarrow {}^sP \\ L_r^t & \xrightarrow{\ d''\ } & L_{r+1}^t \end{array}$$

where $d''$ is the usual densily defined operator which acts componentwise. It is easily verified that ${}^sP\,{}^sP = 0$, ${}^sP d'' = d''\,{}^sP$.

We have defined by (4.2.5) an element ${}_{o}h(s) = ({}_{o}h_1(s), \dots, {}_{o}h_n(s))$ of $L_o^1$. It follows from (4.2.6) that $P_o h = 1$ and from (4.2.7) that every ${}_{o}h_i(s)$ is square integrable with respect to $\delta^N d\lambda$ for $N \geqslant 2n+3$. Similarly ${}_{o}h(s)$ belongs to the domain of $d''$ as

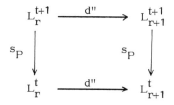

$$d''\left(\frac{\bar z_i - \bar s_i}{|z-s|^2}\right) = -\frac{\bar z_i - \bar s_i}{|z-s|^4}((z_1 - s_1)d\bar z_1 + \ldots + (z_n - s_n)d\bar z_n) + \frac{d\bar z_i}{|z-s|^2}\,.$$

We first define by increasing induction an element ${}_k h(s)$ of the domain of d" in $L_k^{k+1}$ such that

$${}^s P\ {}_k h(s) = d''\ {}_{k-1} h(s)$$

by

$${}_k h(s)_I = \sum_{j=1}^{k+1} (-1)^{k-j+1} \frac{\bar z_{i_j} - \bar s_{i_j}}{|z-s|^2}\, d''_{k-1} h(s)_{(i_1, \ldots, \hat{i}_j, \ldots, i_{k+1})}\,.$$

The system $({}_k h_I)$ is skew symetrical because $P\, d''({}_{k-1}h) = 0$. Moreover, ${}_k h$ belongs to the domain of d" in $L_k^{k+1}$.

As $\Lambda^{n+1}(\mathbf{C}^n) = 0$, we have ${}_n h(s) = 0$. We set ${}_n h'(s) = 0$ and define now, by increasing induction on h, an element ${}_k h'(s)$ of $L_k^{k+2}$ such that

$$d''\ {}_{k-1} h'(s) = {}_k h(s) - {}^s P\ {}_k h'(s).$$

Assume that ${}_n h'(s), \ldots, {}_k h'(s)$ are already defined. We have

$$\int |{}_k h(s) - {}^s P\ {}_k h'(s)|^2\, \delta^N\, d\lambda \leqslant M$$

for some positive integer N and some positive constant M. Moreover

$$\begin{aligned} d''({}_k h(s) - {}^s P\ {}_k h'(s)) &= d''\ {}_k h(s) - {}^s P({}_{k+1} h(s) - {}^s P\ {}_{k+1} h'(s)) \\ &= d''\ {}_k h(s) - {}^s P\ {}_{k+1} h(s) \\ &= 0. \end{aligned}$$

Using Lemma 1, we can find ${}_{k-1} h'(s)$ in $L_k^{k+1}$ such that

$$d''\ {}_{k-1} h'(s) = {}_k h(s) - {}^s P\ {}_k h'(s)$$

and

$$\int |{}_{k-1} h(s)|^2\, \delta_o^4\, \delta^N\, d\lambda \leqslant M.$$

We finally set

$$u(s) = {}_o h(s) - {}^s P\ {}_o h'(s).$$

Obviously u(s) satisfies $d''u(s) = 0$ and ${}^s P u(s) = {}^s P\ {}_o h(s) = 1$ and , as u(s) belongs to $L_o^1$, we have

$$\int |u(s)|^2\, \delta^N(s)\, d\lambda \leqslant M,$$

for some positive integer N and some positive constant M. Then Proposition 2 of

Chapter I shows that $u_1(s),\dots, u_n(s)$ belong to $\mathcal{O}(\delta)$. It is easily seen that all the estimates are independent of s. Thus $u_1(s),\dots, u_n(s)$ are bounded independently of s in $\mathcal{O}(\delta)$.

Remark. Theorem 1 shows that for every pseudoconvex open $\Omega$ in $\mathbf{C}^n$ we can find a positive integer N and a positive number M such that holomorphic functions $u_1(s),\dots, u_n(s)$ can be associated to every point $s \notin \Omega$ so that

$$(z_1-s_1)\,u_1(s)+\dots+(z_n-s_n)\,u_n(s) = 1$$

and

$$\delta_\Omega^N\,|u_i(s)| \leq M, \quad i=1,\dots,n.$$

It is important to note that the estimates used in the proof of Theorem 1 are independent of $\Omega$. Therefore, we can find N, M so that the property is valid for every pseudoconvex open set $\Omega$.

## 4.3.- Spectral functions for z in $\mathcal{O}(\delta)$.

We keep the notations of Section 4.2. A non negative function $\varphi$ on $\mathbf{C}^n$ is spectral for z in $\mathcal{O}(\delta)$ if, for every $s \in \mathbf{C}^n$, we can find holomorphic functions $u_0(s): \zeta \mapsto u_0(s;\zeta),\dots, u_n(s): \zeta \mapsto u_n(s;\zeta)$ in $\Omega$ so that

(4.3.1) $$(z_1-s_1)\,u_1(s)+\dots+(z_n-s_n)\,u_n(s)+\varphi(s)\,u_0(s) = 1$$

and

(4.3.2) $$\delta^N\,|u_i(s)| \leq M, \quad i=0,\dots,n,$$

where N is a positive integer and M a positive constant, both independent of s. Conditions (4.3.1) and (4.3.2) may also be written

(4.3.3) $$(\zeta_1-s_1)u_1(s;\zeta)+\dots+(\zeta_n-s_n)\,u_n(s;\zeta)+\varphi(s)\,u_0(s;\zeta) = 1$$

and

(4.3.4) $$\delta^N(\zeta)\,|u_i(s;\zeta)| \leq M, \; i=0,\dots,n$$

for $s \in \mathbf{C}^n$, $\zeta\in\Omega$.

If $\varphi$ belongs to $\Delta(z;\mathcal{O}(\delta))$, taking $s=\zeta\in\Omega$ in (4.3.3) we have

$$1/\varphi(\zeta) = u_0(\zeta;\zeta).$$

Then, using (4.3.4), we get

$$1/\varphi(\zeta) \leq \sup_{s\in\Omega} |u_0(s;\zeta)| \leq M/\delta^N(\zeta)$$

or

(4.3.5) $$\varphi \geq \gamma \geq 1/M\,\delta^N$$

with

$$(4.3.6) \qquad 1/\gamma = \sup_{s \in \Omega} |u_o(s)| .$$

As $\delta^N |u_o(s)| \leqslant M$, it is easily seen that $u_o(s)$ is locally bounded in $\Omega$ independently of $s$; therefore $-\log \gamma$ is plurisubharmonic (and continuous). In particular $\varphi \geqslant 1/M\ \delta^N$ and every spectral function for $z$ is larger than some function equivalent to $\delta$.

Moreover, if $\delta \in \Delta(z;\mathcal{O}(\delta))$, we get from (4.3.5)

$$\delta \geqslant \gamma \geqslant 1/M \delta^N .$$

Hence $\delta$ is equivalent to a function $\gamma$ such that $-\log \gamma$ is plurisubharmonic in $\Omega$. We shall prove that such a condition is sufficient. We consider a particular case

Theorem 2 (I. Cnop).- Let $\delta$ be a weight function on $\mathbf{C}^n$ bounded by $\delta_o$ and such that $-\log \delta$ is plurisubharmonic in the set $\Omega$ where $\delta$ does not vanish; then $\delta$ is spectral for $z$ in $\mathcal{O}(\delta)$.

We need the following

Lemma 2.- Let $\delta$ be a non negative function on $\mathbf{C}^n$ and $\Omega_1$ denote the subset of $\mathbf{C}^n \oplus \mathbf{C}$ of all $(s,t)$ with $s \in \mathbf{C}^n$, $t \in \mathbf{C}$ such that $|t| < \delta(s)$. Then:

$$\tilde{\delta}(s) = d((s,0), \complement\Omega_1),$$

where $d$ is the distance with respect to the norm $(s,t) \mapsto |s| + |t|$ in $\mathbf{C}^n \oplus \mathbf{C}$. Moreover, assume that $\delta$ is lower semi-continuous, that $\Omega = \{\delta > 0\}$ is pseudoconvex and that $-\log \delta$ is plurisubharmonic in $\Omega$ ; then $\Omega_1$ is pseudoconvex.

Proof. By definition

$$d((s,0), \complement\Omega_1) = \inf_{|t| \geqslant \delta(\zeta)} (|\zeta - s| + |t|)$$

$$= \inf_{\zeta \in \mathbf{C}^n} (|\zeta - s| + \delta(\zeta)) = \tilde{\delta}(s).$$

If $\delta$ is lower semi-continuous, $\Omega$ is open. Obviously, $\Omega_1$ is the subset of $\Omega \times \mathbf{C}$ of all $(s,t)$ such that

$$-\log \delta(s) + \log |t| < 0.$$

If $-\log \delta$ is plurisubharmonic in $\Omega$, then $(s,t) \mapsto -\log \delta(s) + \log |t|$ is plurisubharmonic in $\Omega \times \mathbf{C}$ and Proposition 1 shows that $\Omega_1$ is pseudoconvex.

Proof of Theorem 2. We keep the notations of Lemma 2. As $-\log \delta$ is plurisubharmonic in $\Omega$ and tends to infinity at the boundary, $\Omega$ is pseudoconvex. Then the conditions of Lemma 2 are fulfilled and $\Omega_1$ is pseudoconvex. Using Theorem 1, we see that $\Omega_1$ is spectral for $(z,w)$ in $\mathcal{O}(\delta_{\Omega_1})$ and, using Corollary 1 of Chapter III, that also $\delta_{\Omega_1}$

is spectral for (z, w) in $\mathcal{O}(\delta_{\Omega_1})$, where w denotes the second projection of $\mathbf{C}^n \oplus \mathbf{C}$. Therefore, for every (s, t) in $\mathbf{C}^n \oplus \mathbf{C}$ we can find holomorphic functions $u_i(s,t) : (\zeta, \tau) \mapsto u_i((s,t);(\zeta,\tau))$, $i=0,\dots,n+1$ in $\Omega_1$ such that

$$(\zeta_1 - s_1) u_1((s,t);(\zeta,\tau)) + \dots + (\zeta_n - s_n) u_n((s,t);(\zeta,\tau)) + (\tau - t) u_{n+1}((s,t);(\zeta,\tau))$$
$$+ \delta_{\Omega_1}((s,t)) u_o((s,t);(\zeta,\tau)) = 1$$

and

$$\delta^N_{\Omega_1}((\zeta,\tau)) \, |u_i((s,t);(\zeta,\tau))| \leqslant M,$$

for $(s,t) \in \mathbf{C}^n \oplus \mathbf{C}$ and $(\zeta,\tau) \in \Omega_1$. We set $u_i(s;\zeta) = u_i((s,0);(\zeta,0))$ for $i=0,\dots,n$. Besides, as $\delta$ is a weight function bounded by $\delta_o$, we have $\delta = \mathrm{Min}(\delta_o, \tilde{\delta})$. Using then Lemma 2, we get

$$\delta(s) = \mathrm{Min}\,(\delta_o((s,0)), d((s,0), \complement\Omega_1)) = \delta_{\Omega_1}((s,0)).$$

We have obtained holomorphic functions $u_i(s) : \zeta \mapsto u_i(s;\zeta)$, $i=0,\dots,n$, in $\Omega = \Omega_1 \cap (\mathbf{C}^n \oplus \{0\})$ satisfying

$$(\zeta_1 - s_1) u_1(s;\zeta) + \dots + (\zeta_n - s_n) u_n(s;\zeta) + \delta(s) u_o(s;\zeta) = 1$$

and

$$\delta^N(s) \, |u_i(s;\zeta)| \leqslant M, \quad i=0,\dots,n$$

for $s \in \mathbf{C}^n$, $\zeta \in \Omega$, and the proof of Theorem 2 is complete.

Remark. The positive integer N and the positive number M which appear in the above property are independent of the weight function $\delta$, when the conditions of Theorem 2 are fulfilled.

## 4.4.- Plurisubharmonic regularization

We need a definition. Let $\delta$ be a lower semi-continuous non negative function on $\mathbf{C}^n$ and $\Omega$ the set where $\delta$ does not vanish. We denote by $\hat{\delta}$ the lowest majorant of $\delta$ such that $-\log\hat{\delta}$ is plurisubharmonic in $\Omega$ ; such a function exists: let $\mathcal{E}$ denote the set of all plurisubharmonic functions f in $\Omega$ such that $f \leqslant -\log\delta$ . Taking

$$\varphi(s) = \limsup_{\zeta \to s} \, (\sup_{f \in \mathcal{E}} f(\zeta)),$$

it is well-known that $\varphi$ is plurisubharmonic in $\Omega$ and, as $-\log\delta$ is upper semi-continuous, we have $\varphi \leqslant -\log\delta$ . Then $\hat{\delta} = -\log\varphi$ .

Proposition 2.- Let $\delta$ be a lower semi-continuous non negative function on $\mathbf{C}^n$ and assume that the set $\Omega$ where $\delta$ does not vanish is pseudoconvex. If $-\log\delta$ is plurisubharmonic in $\Omega$ , then also $-\log\hat{\delta}$ is; if $\delta$ satisfies W2, then also $\hat{\delta}$ does.

Proof. This is an easy consequence of Lemma 2. If $-\log\delta$ is plurisubharmonic in $\Omega$, then $\Omega_1$ is pseudoconvex and the function $(s,t)\mapsto -\log d((s,t),\complement\Omega_1)$ is plurisubharmonic. Therefore $-\log\tilde{\hat{\delta}}$ is plurisubharmonic. Further, if $\delta$ satisfies W2, then $\hat{\tilde{\hat{\delta}}}$ is a majorant of $\delta$. But $\hat{\tilde{\hat{\delta}}}$ is plurisubharmonic in $\Omega$ ; hence $\hat{\delta}\leqslant\hat{\tilde{\hat{\delta}}}$. This implies that $\hat{\delta}$ satisfies W2.

Proposition 3.- Let $\delta$ be a weight function on $\mathbf{C}^n$ and assume that the set $\Omega$ where $\delta$ does not vanish is pseudoconvex. Then $\hat{\delta}$, extended by 0 on the complement of $\Omega$, is a weight function and $\mathcal{O}(\hat{\delta}) = \mathcal{O}(\delta)$.

Proof. We note that $\delta_0\geqslant\varepsilon\delta$ for some positive number $\varepsilon$. As $-\log\delta_0$ is plurisubharmonic, we also have $\delta_0\geqslant\varepsilon\hat{\delta}$. As $\hat{\delta}$ satisfies W2, it is a weight function. Clearly $\mathcal{O}(\delta)$ contains $\mathcal{O}(\hat{\delta})$ because $\hat{\delta}\geqslant\delta$. Conversely, let f be holomorphic in $\Omega$ and satisfy $|f|\delta^N\leqslant M$ for some positive integer N and some positive number M. We have

$$1/N\,(\log|f| - \log M)\leqslant -\log\delta$$

and, as $\log|f|$ is plurisubharmonic

$$1/N\,(\log|f| - \log M)\leqslant -\log\hat{\delta}\quad ,$$

that is $|f|\hat{\delta}^N\leqslant M$. Therefore $\mathcal{O}(\delta) = \mathcal{O}(\hat{\delta})$. Moreover $\mathcal{O}(\delta)$ and $\mathcal{O}(\hat{\delta})$ have the same boundedness.

We are now able to characterize the spectrum of z in $\mathcal{O}(\delta)$ when the set where $\delta$ does not vanish is pseudoconvex.

Theorem 3.- Let $\delta$ be a weight function on $\mathbf{C}^n$ and assume that the set $\Omega$ where $\delta$ does not vanish is pseudoconvex. A non negative function $\varphi$ in $\mathbf{C}^n$ is spectral for z in $\mathcal{O}(\delta)$ if and only if $\varphi$ is larger than some function equivalent to $\hat{\delta}$.

Proof. First assume that $\varphi$ is spectral for z in $\mathcal{O}(\delta)$. We have already shown that $\varphi\geqslant\gamma\geqslant 1/M\,\delta^N$ where $-\log\gamma$ is plurisubharmonic in $\Omega$. Further, we easily get $\gamma\geqslant(1/M\,\delta^N)^{\wedge}\geqslant 1/M\,\hat{\delta}^N$.

Conversely, let $\delta_1 = \mathrm{Min}(\hat{\delta},\delta_0)$; then $\mathcal{O}(\delta) = \mathcal{O}(\delta_1)$ and from Theorem 2 we get $\hat{\delta}\in\Delta(z;\mathcal{O}(\delta_1))$.

Corollary 1.- Let $\delta$ be a weight function on $\mathbf{C}^n$ and $\Omega$ denote the set where $\delta$ does not vanish; then $\delta$ is spectral for z in $\mathcal{O}(\delta)$ if and only if $\delta$ is equivalent to some function $\gamma$ such that $-\log\gamma$ is plurisubharmonic in $\Omega$.

The necessary condition has already been proved. If $\delta$ is equivalent to $\gamma$ such that $-\log\gamma$ is plurisubharmonic, then $\hat{\delta}$ is equivalent to $\delta$ and the statement is a consequence of Theorem 3.

A straightforward generalization of Theorem 3 is the following

Proposition 4.- Let $\Delta$ be a directed set of weight functions. Assume that $\mathcal{O}(\Delta)$ is complete and that $\{\delta > 0\}$ is pseudoconvex for each $\delta \in \Delta$. A non negative function $\varphi$ on $\mathbf{C}^n$ is spectral for $z$ in $\mathcal{O}(\Delta)$ if and only if it is larger than a function equivalent to some $\hat{\delta}$ with $\delta \in \Delta$.

We only note that $\varphi$ is spectral for $z$ in $\mathcal{O}(\Delta)$ if and only if there exists some $\delta \in \Delta$ such that $\varphi$ is spectral for $z$ in $\mathcal{O}(\delta)$.

## 4.5.- Domains of holomorphy

An important consequence of the results of the previous section is

Theorem 4.- Let $\Omega$ be a pseudoconvex open set in $\mathbf{C}^n$; there exists a function $f$ of $\mathcal{O}(\delta_\Omega)$ which cannot be holomorphically continued beyond $\Omega$.

Proof. By virtue of Theorem 1, the set $\Omega$ is spectral for $z$ in $\mathcal{O}(\delta_\Omega)$; there exists a positive integer $N$ such that functions $u_1(s), \dots, u_n(s)$ of ${}_N\mathcal{O}(\delta_\Omega)$ can be associated to every point $s \notin \Omega$ so that

$$(4.5.1) \qquad (z_1 - s_1)\, u_1(s) + \dots + (z_n - s_n)\, u_n(s) = 1 .$$

First assume by absurd that there exist a connected open set $\omega$ intersecting $\partial\Omega$ and a connected component $\omega'$ of $\omega \cap \Omega$ such that, for every function $g$ of ${}_N\mathcal{O}(\delta_\Omega)$, we can find some holomorphic function $h$ on $\omega$ which coincides with $g$ on $\omega'$. Let $v_1(s), \dots, v_n(s)$ be holomorphic functions on $\omega$ associated to $u_1(s), \dots, u_n(s)$. By holomorphic continuation (4.5.1) should yield

$$(z_1 - s_1)\, v_1(s) + \dots + (z_n - s_n)\, v_n(s) = 1.$$

Choose now $s \in \omega \cap \partial\Omega$. As $(z_1 - s_1)\, v_1(s) + \dots + (z_n - s_n)\, v_n(s)$ vanishes at $s$, we obtain a contradiction.

Let $(\omega_r)$ denote a denumerable basis of connected open sets intersecting $\partial\Omega$ and for every $r$ let $\omega_{r,1}, \omega_{r,2}, \dots$ denote the sequence of connected components of $\omega_r \cap \Omega$. Clearly each $\mathcal{O}(\omega_r)$ is a Fréchet algebra and each fibrated product

$${}_N\mathcal{O}(\delta_\Omega) \;\pi_{\omega_{r,p}}\; \mathcal{O}(\omega_r) = E_{r,p}$$

is a Fréchet space. We have seen that the first projection

$${}_N\mathcal{O}(\delta_\Omega) \;\pi_{\omega_{r,p}}\; \mathcal{O}(\omega_r) \to {}_N\mathcal{O}(\delta_\Omega),$$

which is obviously continuous and injective, is not onto. The Banach homomorphism theorem shows that the union, when $(r, p)$ varies, of the projections of the $E_{r,p}$ is different from ${}_N\mathcal{O}(\delta_\Omega)$. Let $f$ be a function of ${}_N\mathcal{O}(\delta_\Omega)$ which is contained in the projection of no $E_{r,p}$.

Suppose by absurd that $f$ can be holomorphically continued beyond $\Omega$ : there exist a connected open set $\omega$ intersecting $\partial\Omega$, a connected component $\omega'$ of $\omega\cap\Omega$ and an holomorphic function $g$ on $\omega$ such that $f = g$ on $\omega'$. Clearly $\omega$ intersects $\omega'$ and $\complement\omega'$; as $\omega$ is connected, it intersects $\partial\omega'$. Choose then $s \in \omega\cap\partial\omega'$. We easily see that $s \in \partial\Omega$; then $s$ belongs to some $\omega_r$ which is contained in $\omega$. As $\omega_r \cap \omega'$ is not void, it intersects some $\omega_{r,p}$ and $f = g$ on $\omega_{r,p}$. Thus $f$ belongs to the projection of $E_{r,p}$.

An open domain $\Omega$ of $\mathbf{C}^n$ is called a domain of holomorphy if we cannot find a connected open set $\omega$ intersecting $\partial\Omega$ and a connected component $\omega'$ of $\omega\cap\Omega$ such that for every $f \in \mathcal{O}(\Omega)$ there exists $g \in \mathcal{O}(\omega)$ with $f = g$ on $\omega'$. If $\Omega$ is a domain of holomorphy, it is easily proved that some $f \in \mathcal{O}(\Omega)$ can be found so that no $\omega$, $\omega'$ and $g$ exist with the above properties; it is also proved that $\Omega$ is pseudoconvex. We obtain here the well-known converse statement but Theorem 4 is more precise.

A generalization to weight functions which are not necessarily equal to some $\delta_\Omega$, is given by

Theorem 5.- Let $\delta$ be a weight function on $\mathbf{C}^n$ bounded by $\delta_o$ and such that $-\log\delta$ is plurisubharmonic on the open set $\Omega$ where $\delta$ does not vanish. There exists a family $(f_\alpha)$ of holomorphic functions in $\Omega$ such that

$$1/\delta \leq \sup_\alpha |f_\alpha| \leq M/\delta^N ,$$

for some positive integer $N$ and some positive number $M$, which are independent of $\delta$.

Proof. From Theorem 2, we know that $\delta$ is spectral for $z$ in $\mathcal{O}(\delta)$. Then the statement easily follows from (4.3.5) and (4.3.6) with $\varphi = \delta$.

## 4.6.- Bounded multiplicative linear forms

We now give an easy application of the holomorphic functional calculus.

Theorem 6.- Let $\delta$ be a weight function on $\mathbf{C}^n$ such that the open set $\Omega = \{\delta > 0\}$ is pseudoconvex. For every bounded multiplicative linear form $\chi$ on $\mathcal{O}(\delta)$, there exists some $s \in \Omega$ such that $\chi(f) = f(s)$ for every function $f$ in $\mathcal{O}(\delta)$.

Proof. Upon replacing $\delta$ by $\hat{\delta}$ we may assume that $-\log\delta$ is plurisubharmonic in $\Omega$. Then $\delta$ is spectral for $z$ in $\mathcal{O}(\delta)$. The holomorphic functional calculus at $z$ gives a bounded linear mapping $f \mapsto f[z]$ of $\mathcal{O}(\delta)$ into $\mathcal{O}(\delta)$. For each $s \in \Omega$, let

${}^{s}\chi$ be the bounded multiplicative linear form $f \mapsto f(s)$ on $\mathcal{O}(\delta)$. From Proposition 11 of Chapter III we get, for every function f in $\mathcal{O}(\delta)$,

$$ {}^{s}\chi\,(f[z]) = f({}^{s}\chi(z_1),\ldots,{}^{s}\chi(z_n)), $$

that is $f[z](s) = f(s)$. Therefore $f[z] = f$ and the holomorphic functional calculus is the identity mapping of $\mathcal{O}(\delta)$. Now let $\chi$ be a bounded multiplicative linear form on $\mathcal{O}(\delta)$. Proposition 11 of Chapter III ensures that $s = (\chi(z_1),\ldots,\chi(z_n))$ belongs to $\Omega$ and that $\chi(f[z]) = f(s)$ for every function f in $\mathcal{O}(\delta)$. Then $\chi(f) = f(s)$ and the proof of the statement is complete.

Using Theorem 6, we immediately obtain

Proposition 5.- Let $\Delta$ be a directed set of weight functions such that for each $\delta \in \Delta$, the set $\{\delta > 0\}$ is equal to some fixed pseudoconvex open set $\Omega$. For every bounded multiplicative linear form $\chi$ on $\mathcal{O}(\Delta)$, there exists some $s \in \Omega$ such that $\chi(f) = f(s)$ for every function f in $\mathcal{O}(\Delta)$.

As a consequence, we give another characterization of pseudoconvex domains.

Proposition 6.- Let $\Omega$ be an open set in $\mathbf{C}^n$; then $\Omega$ is pseudoconvex if and only if $\Omega$ every bounded multiplicative linear form on $\mathcal{O}(\Omega)$ is equal to some ${}^{s}\chi : f \mapsto f(s)$ with $s \in \Omega$.

Proof. If $\Omega$ is pseudoconvex, we only have to apply Proposition 5. Conversely, if is not pseudoconvex, there exists a compact set K in $\Omega$ such that $\widehat{K}_\Omega$ contains a sequence $(t_p)$ which converges to some boundary point t of $\Omega$. Let $\mathcal{U}$ be an ultrafilter on $\mathbf{N}$ which converges to infinity. If f belongs to a bounded subset of $\mathcal{O}(\Omega)$, there exists some positive number M such that $|f(\zeta)| \leq M$ for $\zeta \in K$. Therefore, as $|f|$ is log-plurisubharmonic, by definition of $\widehat{K}_\Omega$ we have $|f(t_p)| \leq M$. Thus we can set

$$ \chi(f) = \lim_{\mathcal{U}} f(t_p). $$

Moreover $|\chi(f)| \leq M$. Hence $\chi$ is a bounded multiplicative linear form. As $\chi(z) = t$ we see that $\chi$ is not equal to some ${}^{s}\chi$ with $s \in \Omega$.

Notes

The equivalent definitions of pseudoconvex domains are well-known (see H.J. Bremermann (1)). The fact that a pseudoconvex open set $\Omega$ is spectral for z in $\mathcal{O}(\delta_\Omega)$ is a particular case, but with parameter, of a result of L. Hörmander (3) mentioned as Corollary 1 of Section 5.6; it is the case when $f_1 = z_1 - s_1, \ldots, f_n = z_n - s_n$. The proof is similar and the same double complex $L_r^t$ is used. It depends on the famous estimates of L. Hörmander for the d"-operator (see L. Hörmander (1), (2) and also J.B. Poly (1)). The increasing and decreasing inductions are taken from the original proof of Theorem 2, or Corollary 1 of Section 4.4, by I. Cnop (1), (2), (3). Here the study of spectral functions is deduced from that of spectral sets. The plurisubharmonic regularization and the other results of Section 4.4 are due to I. Cnop and the author (1). The fact that pseudoconvex domains and domains of holomorphy are the same is a basic result of complex analysis in $\mathbf{C}^n$ (see B.J. Bremermann (1), F. Norguet (1), K. Oka (2)). The existence for every boundary point of a pseudoconvex domain of an holomorphic function with polynomial growth which is singular at this point is due to R. Narashiman (1), where this domain is bounded. Theorem 4 for a bounded domain is given by N. Sibony (1). Theorem 6 and Proposition 5 are direct applications of the holomorphic functional calculus of b-algebras, whereas Proposition 6 is well-known (see R.C. Gunning and H. Rossi (1)).

# CHAPTER V

## DECOMPOSITION PROPERTY FOR ALGEBRAS OF HOLOMORPHIC FUNCTIONS

We introduce for polynormed algebras H of holomorphic functions on a given domain $\Omega$ , the following property: every $f \in H$ vanishing at $s \in \Omega$ can be written $(z_1 - s_1) u_1 + \ldots + (z_n - s_n) u_n$, so that $u_1, \ldots, u_n$ are bounded in H when f varies in a bounded subset of H and s in $\Omega$. Such a decomposition property is proved for an algebra $\mathcal{O}(\delta)$ when $\delta$ is a weight function and the open set $\{\delta > 0\}$ is pseudoconvex, and for a subalgebra $\mathcal{O}(\delta')$ of $\mathcal{O}(\delta)$ when $\delta, \delta'$ are weight functions and the open set $\{\delta' > 0\}$ is pseudoconvex. We also replace $\mathcal{O}(\delta)$, $\mathcal{O}(\delta')$ by inductive limits $\mathcal{O}(\Delta)$, $\mathcal{O}(\Delta')$. For a general H satisfying the decomposition property, we describe the restriction to $\Omega$ of the spectrum of z in $\widehat{H}$ and discuss the structure of H when a condition of convexity of $\Omega$ with respect to H is fulfilled. We apply these results to the study of b-ideals and finitely generated ideals of algebras $\mathcal{O}(\delta)$.

### 5.1.- Preliminaries

Let $\Omega$ be an open set in $\mathbf{C}^n$ and H be a subalgebra of $\mathcal{O}(\Omega)$ containing the polynomials, equipped with an algebra boundedness. We say that H has the decomposition property over $\Omega$ if the following condition is fulfilled:

For every function f in H and every point s in $\Omega$ , with $f(s) = 0$, one can write

$$f = (z_1 - s_1) u_1 + \ldots + (z_n - s_n) u_n ,$$

so that $u_1, \ldots, u_n$ are bounded in H, when f varies in a bounded subset of H and s in $\Omega$.

It is equivalent to ask that for every bounded set B in H, there exists another bounded set B' such that every function f in B vanishing at $s \in \Omega$ belongs to $(z_1 - s_1) B' + \ldots + (z_n - s_n) B'$.

Note that the property depends on the boundedness of H. We give a less restrictive condition which is actually equivalent to the decomposition property.

Proposition 1.- Assume that for every $f \in H$ and $s \in \Omega$, with $f(s) = 0$, one can write

$$f = (z_1 - s_1)\, {}^s u_1 + \ldots + (z_n - s_n)\, {}^s u_n ,$$

so that $\delta_o^N(s)\, {}^s u_1, \ldots, \delta_o^N(s)\, {}^s u_n$ are bounded in H for some integer N, when f varies in a bounded subset of H and s in $\Omega$. Then the same property holds for $N = -1$ and H has the decomposition property over $\Omega$.

Proof. We consider the coefficients

$${}^s U_i = -\bar{s}_i\, \delta_o^2(s) , \quad i = 1, \ldots, n,$$

and

$${}^s Y = (1 + \langle z, \bar{s} \rangle)\, \delta_o^2(s).$$

They obviously satisfy

$$\langle z - s, {}^s U \rangle + {}^s Y = 1.$$

Now let N be a non negative integer such that

$$f = \langle z - s, {}^s u \rangle ,$$

where f and $\delta_o^N(s)\, {}^s u_1, \ldots, \delta_o^N(s)\, {}^s u_n$ vary in a bounded subset of H. We also have

$$f = \langle z - s, {}^s U \rangle f + {}^s Y f$$
$$= \langle z - s, {}^s U + {}^s Y\, {}^s u \rangle ,$$

and each $\delta_o^{N-1}(s) ({}^s U_i + {}^s Y\, {}^s u_i)$ varies in a bounded subset of H. By decreasing induction on N, the statement is therefore proved.

Corollary 1.- The algebra $\mathcal{O}(\delta_o)$ of polynomials has the decomposition property over $\mathbf{C}^n$.

Proof. First let p be a polynomial such that $p(0) = 0$. One can write

$$p = z_1 q_1 + \ldots + z_n q_n ,$$

where $q_1, \ldots, q_n$ are polynomials and $q_i$ only depends on $z_1, \ldots, z_i$. Moreover such a decomposition is unique and $q_1, \ldots, q_n$ are bounded in $\mathcal{O}(\delta_o)$ when p is; if r is a bound for the degree of p, the degree of each $q_i$ is bounded by $r - 1$.

Assume now that $p(s) = 0$; we may consider the polynomial ${}^s p = p(z+s)$ and write as above

$$^{s}p = z_1 \, {}^{s}q_1 + \ldots + z_n \, {}^{s}q_n \, .$$

Thus

$$p = (z_1 - s_1) \, {}^{s}q_1(z-s) + \ldots + (z_n - s_n) \, {}^{s}q_n(z-s) \, .$$

If $p$ is bounded in $\mathcal{O}(\delta_o)$ and $r$ is a bound for the degree of $p$, then $\delta_o(s)^r . {}^{s}p$ is bounded in $\mathcal{O}(\delta_o)$; therefore each $\delta_o(s)^r \, . \, {}^{s}q_i$ is bounded and so each $\delta_o(s)^{2r-1} . \, {}^{s}q_i(z-s)$ is.

Corollary 2.- Assume that each derivative $\partial / \partial z_j$ is a bounded linear mapping of $H$ into $H$ and let $r$ be a non negative integer. In order that $H$ has the decomposition property, it suffices that for every $f \in H$ vanishing at $s$ along with its derivatives up to order $r$, one can write $f = (z_1 - s_1) u_1 + \ldots + (z_n - s_n) u_n$, so that $u_1, \ldots, u_n$ are bounded in $H$, when $f$ varies in a bounded subset of $H$ and $s$ in $\Omega$ .

Proof. Suppose that $f$ vanishes at $s$ and belongs to some fixed bounded subset of $H$. Writing

$$^{s}g = f - f' . (z-s) + \ldots + (-1)^r f^{(r)} . (z-s)^r \, ,$$

as ${}^{s}g(s) = 0$ and ${}^{s}g' = f^{(r+1)} . (z-s)^r$, clearly ${}^{s}g$ vanishes at $s$ along with its derivatives up to order $r$. Moreover $\delta_o^r(s) \, {}^{s}g$ is bounded in $H$ and the hypothesis implies

$$\delta_o^r(s) \, {}^{s}g = \langle z-s, {}^{s}u \rangle \, ,$$

where ${}^{s}u_1, \ldots, {}^{s}u_n$ belong to some fixed bounded subset of $H$. Finally

$$f = \langle z-s, f' \rangle + \ldots + (-1)^{r+1} \langle z-s, f^{(r)} . (z-s)^{r-1} \rangle + \langle z-s, \delta_o^{-r}(s) \, {}^{s}u \rangle \, ,$$

and the statement follows from Proposition 1.

If $H$ is a complete subalgebra with decomposition property, for every $s \in \Omega$ , the ideal ${}^{s}\mathfrak{J}$ of all functions $f$ in $H$ vanishing at $s$ is equal to the b-ideal $\mathrm{idl}(z-s; H)$ generated by $z_1 - s_1, \ldots, z_n - s_n$. Moreover the ideal $\mathfrak{J}$ of $\mathcal{C}(\delta_o; H)$ of all families $({}^{s}f)_{s \in \Omega}$ such that ${}^{s}f(s) = 0$ is equal to the b-ideal $\mathrm{idl}(z-s; \mathcal{C}(\delta_o; H))$ : if a family $({}^{s}f)$ belongs to a bounded subset of $\mathfrak{J}$ , there exists a positive integer $N$ such that $\delta_o^N(s) \, {}^{s}f$ lies in some fixed bounded set in $H$ and

$$\delta_o^N(s) \, {}^{s}f = \langle z-s, {}^{s}u \rangle \, ,$$

where ${}^{s}u_1, \ldots, {}^{s}u_n$ also lie in some fixed bounded set in $H$. Then as

$$^{s}f = \langle z-s, \delta_o^{-N}(s) \, {}^{s}u \rangle \, ,$$

clearly ${}^{s}f$ is bounded in $\mathrm{idl}(z-s; \mathcal{C}(\delta_o; H))$. Conversely $\mathfrak{J} = \mathrm{idl}(z-s; \mathcal{C}(\delta_o; H))$ implies the decomposition property.

5.2.- Decomposition property for $\mathcal{O}(\delta)$.

The fact that $\mathcal{O}(\delta_o)$ has the decomposition property is nothing but a particular case of the following result:

Theorem 1.- Let $\delta$ be a weight function on $\mathbf{C}^n$ and assume that the open set $\Omega$ where $\delta$ does not vanish is pseudoconvex. Then $\mathcal{O}(\delta)$ has the decomposition property over $\Omega$.

Proof. Upon replacing $\delta$ by the weight function $\hat{\delta}$ defined in Section 4.4, we may assume that $-\log\delta$ is plurisubharmonic in $\Omega$. We introduce another variable; now $s = (s_1, \ldots, s_n)$ denotes the first projection of $\mathbf{C}^n \times \mathbf{C}^n$ and $z = (z_1, \ldots, z_n)$ the second one. Using Theorem 2 of Chapter IV and identifying $\mathcal{O}(\delta)$ with the b-algebra of all functions in $\mathcal{O}(\delta \otimes \delta)$ which only depend on $z$, we know that $\delta$ is spectral for $z$ in $\mathcal{O}(\delta \otimes \delta)$. Applying then the holomorphic functional calculus, we define for every function $f$ in $\mathcal{O}(\delta)$ an element $f[z]$ of $\mathcal{O}(\delta \otimes \delta)$. Let $(\sigma, \zeta)$ be an element of $\Omega \times \Omega$ and $\chi$ be the bounded multiplicative linear form $f \mapsto f(\sigma, \zeta)$. From Proposition 11 of Chapter III, we have $\chi(f[z]) = f(\chi(z))$, that is $f[z](\sigma,\zeta) = f(\zeta)$. Hence $f[z] = f(z)$. Similarly $\delta$ is spectral for $s$ in $\mathcal{O}(\delta \otimes \delta)$ and $f[s]$ is defined and satisfies $f[s] = f(s)$. But $z \equiv s$ modulo the b-ideal $I = \mathrm{idl}(z-s; \mathcal{O}(\delta \otimes \delta))$ and $f[z] - f[s]$ is bounded in $I$ when $f$ varies in a bounded subset of $\mathcal{O}(\delta)$. We therefore can find $g_1, \ldots, g_n$ in a bounded subset of $\mathcal{O}(\delta \otimes \delta)$ such that

$$f(z) - f(s) = (z_1 - s_1)g_1 + \ldots + (z_n - s_n)g_n.$$

Returning to our previous notations, let $s$ denote an arbitrary point in $\mathbf{C}^n$, let ${}^s g_i = g_i(s, z)$ and assume that $f(s) = 0$. Then

$$f = \langle z - s, {}^s g \rangle \quad . \tag{5.2.1}$$

Obviously ${}^s g_1, \ldots, {}^s g_n$ are in $\mathcal{O}(\delta)$; they are not bounded when $s$ varies but there exists a positive integer $N$ such that each $\delta^N(s)\, {}^s g_i$ is bounded in $\mathcal{O}(\delta)$. Using again the fact that $\delta$ and thereby $\delta^N$ are spectral for $z$ in $\mathcal{O}(\delta)$, we obtain coefficients ${}^s u_o, {}^s u_1, \ldots, {}^s u_n$, bounded independently of $s$ in $\mathcal{O}(\delta)$ such that

$$\langle z - s, {}^s u \rangle + \delta^N(s)\, {}^s u_o = 1 . \tag{5.2.2}$$

Multiplying (5.2.2) by $f$ and using (5.2.1), we get

$$f = \langle z - s, {}^s u\, f \rangle + \langle z - s, {}^s u_o\, \delta^N(s)\, {}^s g \rangle ,$$

and the decomposition property is now proved.

As an immediate consequence of Theorem 1, we obtain the following more general statement:

Proposition 2.- Let $\Delta$ be a directed set of weight functions such that for each $\delta \in \Delta$, the set $\{\delta > 0\}$ is equal to some fixed pseudoconvex open set $\Omega$. Then $\mathcal{O}(\Delta)$ has the decomposition property over $\Omega$.

For instance, when $\Omega$ is pseudoconvex, the algebra $\mathcal{O}(\Omega)$ has the decomposition property; for every convex increasing mapping $\varphi$ of $[0, \infty[$ into $[0, \infty[$, $\mathcal{O}(\Lambda_{\Omega,\varphi})$ also has.

5.3.- Decomposition property for subalgebras

Let $\delta, \delta'$ be weight functions on $\mathbf{C}^n$ such that $\delta'$ is larger than some function equivalent to $\delta$ and $\Omega'$ denote the set where $\delta'$ does not vanish.

Theorem 2.- If $\Omega'$ is pseudoconvex, the algebra $\mathcal{O}(\delta')$, equipped with the boundedness induced by $\mathcal{O}(\delta)$, has the decomposition property over $\Omega'$.

We need the following

Lemma 1.- Let f be an holomorphic function defined on a pseudoconvex open set $\omega$ and

$$\delta_f(\zeta) = \mathrm{Min}\ (\inf_{\eta \in \mathbf{C}^n} |\zeta - \eta| + 1/|f(\eta)|\ ,\ \delta_o(\zeta)),$$

with $1/|f(\eta)| = 0$ if $\eta \notin \omega$. Then $|f|\delta_f \leqslant 1$ and if f vanishes at $s \in \omega$, there exist holomorphic functions $u_1, \ldots, u_n$ in $\omega$ such that

$$f = (z_1 - s_1)\, u_1 + \ldots + (z_n - s_n)\, u_n$$

and

$$|u_i|\, \delta_f^{N_o} \leqslant M_o\ , \quad i = 1, \ldots, n$$

where $N_o$, $M_o$ only depend on n.

Proof. We note that $\delta_f$ is the weight function $\tilde{\varphi}_f$, where $\varphi_f$ is defined by

$$\varphi_f = \mathrm{Min}\ (1/|f|\ ,\ \delta_o).$$

Obviously $|f|\varphi_f \leqslant 1$ and $|f|\delta_f \leqslant 1$; thus f belongs to $\mathcal{O}(\delta_f)$ and satisfies in $\mathcal{O}(\delta_f)$ an estimate independent of f.

As $-\log \varphi_f$ is plurisubharmonic in $\omega$, Proposition 2 of Chapter IV shows that also $-\log \delta_f$ is. If f vanishes at $s \in \omega$, applying Theorem 1 in $\mathcal{O}(\delta_f)$ we obtain holomorphic functions $u_1, \ldots, u_n$ on $\omega$ such that the conclusions of Lemma 1 hold. It is clear that $N_o$ and $M_o$ only depend on n, because the constants appearing in the property $\delta_f \in \Delta(z;\ \mathcal{O}(\delta_f))$ and the holomorphic functional calculus $\mathcal{O}(\delta_f) \to \mathcal{O}(\delta_f \otimes \delta_f)$, only depend on n.

Proof of Theorem 2. Let f be a function of $\mathcal{O}(\delta')$ vanishing at $s \in \Omega'$ and assume that f belongs to a bounded subset B of $\mathcal{O}(\delta)$: there exist a positive integer N and a positive number M such that

$$|f|\delta^N \leqslant M. \tag{5.3.1}$$

Using Lemma 1, we can write $f = \langle z-s, u\rangle$ where $u_1,\dots,u_n$ are holomorphic in $\Omega'$ and satisfy

$$\delta_f^{N_o}\,|u_i| \leqslant M_o, \quad i=1,\dots,n. \tag{5.3.2}$$

From (5.3.1), we get $1/|f| \geqslant \varepsilon\delta^N$ for $\varepsilon > 0$ small enough and we may assume that $\varepsilon\delta^N$ is bounded by $\delta_o$ and satisfies W2. Then $\varphi_f \geqslant \varepsilon\delta^N$ and also $\delta_f \geqslant \varepsilon\delta^N$. Thus (5.3.2) implies

$$\delta^{NN_o}\,|u_i| \leqslant M_o\,\varepsilon^{-N_o}$$

and $u_1,\dots,u_n$ are bounded in $\mathcal{O}(\delta)$ when f varies in B. Reasoning similar, from $\delta'^{N'}|f| \leqslant M'$ for some $N', M'$ depending on f, we deduce that

$$\delta'^{N'N_o}\,|u_i| \leqslant M_o\,\varepsilon'^{-N_o}$$

for some $\varepsilon' > 0$ also depending on f. Hence $u_1,\dots,u_n$ belong to $\mathcal{O}(\delta')$.

More generally, we may consider directed sets $\Delta$, $\Delta'$ of weight functions such that each $\delta' \in \Delta'$ is larger than a function equivalent to some $\delta \in \Delta$. Then

Theorem 3.- Assume that each $\{\delta' > 0\}$, with $\delta' \in \Delta'$, is pseudoconvex and contains a fixed open set $\Omega$. Then $\mathcal{O}(\Delta')$, equipped with the boundedness induced by $\mathcal{O}(\Delta)$, has the decomposition property over $\Omega$, if $\Omega \supset \{\delta > 0\}$ for each $\delta \in \Delta$.

For instance, let $\Delta$ be a directed set of weight functions such that for each $\delta \in \Delta$, the set $\{\delta > 0\}$ is equal to some fixed pseudoconvex open set $\Omega$; let $K, K'$ be compact subsets of $\Omega$ such that $K'$ is a neighbourhood of K. Every function $f \in \mathcal{O}(\Delta)$ vanishing at $s \in \Omega$ can be written

$$f = (z_1 - s_1)\,g_1 + \dots + (z_n - s_n)\,g_n,$$

so that $g_1,\dots,g_n$ belong to $\mathcal{O}(\Delta)$ and are uniformly bounded on K when f is uniformly bounded on $K'$. To prove such a property, we may consider on $\mathcal{O}(\Delta)$ the structure induced by $\mathcal{O}(\delta_{\mathring{K}'})$ and apply Theorem 3. If f is uniformly bounded on $K'$, it is bounded in $\mathcal{O}(\delta_{\mathring{K}'})$; we therefore can find $g_1,\dots,g_n$ in $\mathcal{O}(\Delta)$ so that they are bounded in $\mathcal{O}(\delta_{\mathring{K}'})$. But $\delta_{\mathring{K}'}$ is uniformly bounded from below on K and $g_1,\dots,g_n$ are uniformly bounded on K.

5.4.- Spectral functions for z

We consider an open set $\Omega$ in $\mathbf{C}^n$ and a subalgebra H of $\mathcal{O}(\Omega)$ containing the polynomials, equipped with an algebra boundedness. We assume that the structure of H is finer than the structure induced by $\mathcal{O}(\Omega)$ and that H has the decomposition property over $\Omega$. Such an algebra will be called a subalgebra with decomposition property of $\mathcal{O}(\Omega)$

As H may not be complete, we introduce the b-algebra $\hat{H}$. For every point s in $\Omega$ the multiplicative linear form ${}^s\chi : f \mapsto f(s)$ is a bounded linear form on H; therefore ${}^s\chi$ can be continued as a bounded multiplicative linear form on $\hat{H}$. As ${}^s\chi(f) = {}^s\chi(g)$ implies $f = g$ when in H, the natural homomorphism $H \to \hat{H}$ is injective.

Let $s \in \Omega$ and B be an absolutely convex bounded subset of H such that $1 \in E_B$; we denote by ${}^s\mathfrak{J}$ the ideal of H composed of all functions f such that $f(s) = 0$ and by $\delta_B(s)$ the distance, in the pseudonormed vector space $E_B$, from 1 to the linear space ${}^s\mathfrak{J} \cap E_B$.

Proposition 3.- Let H be a subalgebra with decomposition property of $\mathcal{O}(\Omega)$; the set of all functions $\delta_B$ is a basis of the restriction to $\Omega$ of the spectrum of z in $\hat{H}$.

Proof. We first show that every function $\delta_B$ is the restriction to $\Omega$ of a spectral function for z in $\hat{H}$. Choose $\varepsilon \in ]0, 1]$ and let $s \in \Omega$; by definition of $\delta_B(s)$, there exists some ${}^s_\varepsilon u_o$ in B such that

$$1 - (\delta_B(s) + \varepsilon)\, {}^s_\varepsilon u_o = {}^s_\varepsilon f$$

belongs to ${}^s\mathfrak{J}$. Clearly ${}^s_\varepsilon f$ is bounded in H as $\delta_B(s)$ is bounded by $\|1\|_B$. Using the decomposition property, we write

$${}^s_\varepsilon f = \langle z - s, {}^s_\varepsilon u \rangle ,$$

where ${}^s_\varepsilon u_1, \ldots, {}^s_\varepsilon u_n$ belong to an absolutely convex bounded subset B' of H which does not depend on $s \in \Omega$ and $\varepsilon > 0$. Thus

$$\langle z - s, {}^s_\varepsilon u \rangle + \delta_B(s)\, {}^s_\varepsilon u_o = 1 - \varepsilon\, {}^s_\varepsilon u_o .$$

Applying now Proposition 5 of Chapter II and using an argument similar to that of Proposition 4 of Chapter III, we can find an absolutely convex bounded subset B'' of $\hat{H}$ such that 1 belongs to

$$(z_1 - s_1)B'' + \ldots + (z_n - s_n)B'' + \delta_B(s)\, B'' ,$$

for every $s \in \Omega$. Therefore $\delta_B$ is the restriction to $\Omega$ of some function in $\Delta(z; \hat{H})$.

Conversely, let $\delta \in \Delta(z; \hat{H})$. We can find coefficients ${}^s u_o, \ldots, {}^s u_n$, bounded in $\hat{H}$ independently of s and such that

$$\langle z - s, {}^s u \rangle + \delta(s)\, {}^s u_o = 1.$$

Let B' be an absolutely convex bounded subset of H such that each ${}^s u_i$ belongs to the unit ball of $\widehat{E}_{B'}$ : for every $\varepsilon > 0$, there exist ${}^s_\varepsilon u_o, {}^s_\varepsilon u_1, \dots, {}^s_\varepsilon u_n$ in B such that

$$\| {}^s_\varepsilon u_i - {}^s u_i \|_{\widehat{E}_{B'}} \leqslant \varepsilon \quad ,$$

for $i = 0, \dots, n$. Choose another absolutely convex bounded subset B of H such that B contains $B' \cup z_1 B' \cup \dots \cup z_n B'$ ; each multiplication by $z_i$ continuously maps $E_{B'}$ into $E_B$, and therefore $\widehat{E}_{B'}$ into $\widehat{E}_B$. In $\widehat{E}_B$, for s fixed in $\Omega$ , we get

$$\lim_{\varepsilon \to 0} \left[ \langle z - s, {}^s_\varepsilon u \rangle + \delta(s)\, {}^s_\varepsilon u_o \right] = 1 .$$

But $\langle z - s, {}^s_\varepsilon u \rangle$ belongs to ${}^s\mathfrak{J}$ and ${}^s_\varepsilon u_o$ to B. Hence $\delta(s) \geqslant \delta_B(s)$ and the restriction of $\delta$ to $\Omega$ is larger than some function $\delta_B$.

We now give another characterization of $\delta_B$.

Proposition 4.- <u>Let</u> B <u>be an absolutely convex bounded set of complex functions defined on</u> $\Omega$ ; <u>then</u>

$$1/\delta_B = \sup_{f \in B} |f| .$$

<u>Proof</u>. Consider a function f in B and a point s in $\Omega$ . Writing $f = f(s) + g$, we define some g in ${}^s\mathfrak{J}$ . By definition of $\delta_B(s)$, there exists a linear form $\mu$ on $E_B$ vanishing on ${}^s\mathfrak{J}$ and such that $\|\mu\| \leqslant 1$, $\mu(1) = \delta_B(s)$. Then $\mu(f) = f(s)\,\mu(1)$ and

$$|f(s)|\, \delta_B(s) \leqslant 1 .$$

Taking the supremum on $f \in B$, we obtain

$$1/\delta_B \geqslant \sup_{f \in B} |f(s)| .$$

To prove the converse inequality, we set

$$1/\delta = \sup_{f \in B} |f| .$$

Obviously B is contained in the set of all complex functions f on $\Omega$ such that $|f|\,\delta \leqslant 1$. Therefore the Minkowski functional of B is larger than the pseudonorm

$$f \mapsto \sup_{\zeta \in \Omega} |f(\zeta)|\, \delta(\zeta)$$

and

$$\delta_B(s) \geqslant \inf_{f \in {}^s\mathfrak{J}} \sup_{\zeta \in \Omega} |1 - f(\zeta)|\, \delta(\zeta).$$

Hence, taking $\zeta = s$, we get

$$\delta_B(s) \geqslant \delta(s).$$

From Propositions 3 and 4, we immediately obtain

Theorem 4.- Let H be a subalgebra with decomposition property of $\mathcal{O}(\Omega)$; a basis of the restriction to $\Omega$ of the spectrum of z in $\widehat{H}$ is composed of all functions $1/\sup_{f\in B} |f|$, where B is a bounded set in H.

5.5.- Convexity with respect to algebras of holomorphic functions

Keeping the notations of the preceding section, we discuss convexity of $\Omega$ with respect to the subalgebra H.

Theorem 5.- Let H be a subalgebra with decomposition property of $\mathcal{O}(\Omega)$; the following properties are equivalent :

(i) There exists a bounded subset B of H such that for every boundary point s of $\Omega$, the function $\sup_{f\in B} |f|$ is not bounded in any component of $\Omega$ near s.

(ii) $\Omega$ is pseudoconvex and $\mathcal{O}(\delta_\Omega)$ is a subalgebra of $\widehat{H}$ with a finer structure.

(iii) $\Omega$ is spectral for z in $\widehat{H}$.

Proof. First assume that (i) holds. Then for every connected open set $\omega$ intersecting $\partial\Omega$ and every component $\omega'$ of $\omega\cap\Omega$, the projection $\mathcal{O}(\Omega)\,\pi_{\omega'}\mathcal{O}(\omega) \to \mathcal{O}(\Omega)$ cannot be onto; if it was, by virtue of the Banach homomorphism theorem, it would be an isomorphism of Fréchet algebras. As B is bounded in $\mathcal{O}(\Omega)$, it would also be bounded in $\mathcal{O}(\omega)$ and $\sup_{f\in B} |f|$ would be bounded in the neighbourhood of every point s of $\omega\cap\partial\omega'$ in $\omega'$. Hence $\Omega$ is a domain of holomorphy and thereby pseudoconvex.

Theorem 4 shows that there exists some weight function $\delta$ in $\Delta(z;\widehat{H})$ such that $\delta \leqslant 1/\sup_{f\in B}|f|$ on $\Omega$. Being continuous, $\delta$ vanishes on the boundary of $\Omega$. Extending functions by 0 on the complement of $\Omega$, we obtain a bounded linear mapping of $\mathcal{O}(\delta_\Omega)$ into $\mathcal{O}(\delta)$. Besides, the holomorphic functional calculus at z yields a bounded linear mapping $\mathcal{O}(\delta) \to \widehat{H}$. We have already noted that for every $s\in\Omega$, the bounded multiplicative linear form ${}^s\chi : f \mapsto f(s)$ can be continued to $\widehat{H}$; moreover, from Proposition 11 of Chapter III, we have for every f in $\mathcal{O}(\delta)$

$$ {}^s\chi\,(f[z]) = f({}^s\chi\,(z)) = f(s). $$

Hence the bounded linear mapping $\mathcal{O}(\delta_\Omega) \to \widehat{H}$ is injective and multiplicative; we may identify $\mathcal{O}(\delta_\Omega)$ with a subalgebra of $\widehat{H}$ with a finer boundedness and condition (ii) follows.

Obviously (ii) implies (iii) because Theorem 1 of Chapter IV shows that $\Omega$ is spectral for z in $\mathcal{O}(\delta_\Omega)$. Finally (iii) implies that $\delta_\Omega$ is spectral for z in $\widehat{H}$. Using Theorem 4, we can find a bounded subset B of H such that $1/\delta_\Omega \leqslant \sup_{f\in B} |f|$ and condition (i) holds.

We set a definition.

Definition 1.- Let H be a subalgebra with decomposition property of $\mathcal{O}(\Omega)$; if the equivalent conditions of Theorem 5 are fulfilled, we say that $\Omega$ is convex with respect to H, or H-convex.

If $\Omega$ is H-convex, it is pseudoconvex. Conversely a pseudoconvex $\Omega$ is convex with respect to $\mathcal{O}(\delta_\Omega)$, $\mathcal{O}(\Omega)$ or more generally every $\mathcal{O}(\Delta)$ where $\Delta$ is a directed set of weight functions such that $\Omega = \{\delta > 0\}$ for each $\delta \in \Delta$.

Let H be a subalgebra of $\mathcal{O}(\Omega)$ such that $\Omega$ is H-convex. As an immediate consequence of Proposition 3 and Theorem 5 (iii), a basis of the spectrum of z in $\hat{H}$ is the set $\Delta$ of all weight functions $\mathrm{Min}(\delta_\Omega, \tilde{\delta}_B)$; as $\Omega$ is pseudoconvex and $-\log \delta_B = \sup_{f \in B} \log |f|$ is plurisubharmonic in $\Omega$, each $-\log \mathrm{Min}(\delta_\Omega, \tilde{\delta}_B)$ is also plurisubharmonic in $\Omega$.

Every absolutely convex bounded set B in H such that $1 \in E_B$ is bounded in $\mathcal{O}(\mathrm{Min}(\delta_\Omega, \tilde{\delta}_B))$; therefore the identity is a bounded linear mapping of H into $\mathcal{O}(\Delta)$. Further the holomorphic functional calculus yields a bounded linear mapping $\mathcal{O}(\Delta) \to \hat{H}$ such that $H \to \mathcal{O}(\Delta) \to \hat{H}$ is the identity mapping.

Proposition 5.- Let $\Omega$ be an open set in $\mathbf{C}^n$ and H a subalgebra of $\mathcal{O}(\Omega)$ such that $\Omega$ is H-convex. Every bounded multiplicative linear mapping $\chi$ on H is equal to some ${}^s\chi : f \mapsto f(s)$ with $s \in \Omega$.

We note that $\chi$ can be continued to $\hat{H}$. In view of Proposition 11 of Chapter III, we have $s = \chi(z) \in \Omega$ and $\chi(f[z]) = f(\chi(z)) = f(s)$ for every function f in $\mathcal{O}(\Delta)$. As $f[z] = f$ when f belongs to H, the statement is proved.

When $H = \hat{H}$, we get

Theorem 6.- Let H be a complete subalgebra with decomposition property of $\mathcal{O}(\Omega)$. If $\Omega$ is H-convex, then $H = \mathcal{O}(\Delta)$, where $\Delta$ consists of all weight functions $\mathrm{Min}(\delta_\Omega, \tilde{\delta}_B)$.

Let $\Omega$ denote a pseudoconvex open set in $\mathbf{C}^n$; complete subalgebras with decomposition property H of $\mathcal{O}(\Omega)$ such that $\Omega$ is H-convex are exactly algebras $\mathcal{O}(\Delta)$, where $\Delta$ is a directed set of weight functions and $\Omega$ is the set where each $\delta \in \Delta$ does not vanish.

## 5.6.- Ideals of holomorphic functions

We consider in this section an open set $\Omega$ in $\mathbf{C}^n$ and a complete subalgebra with decomposition property H of $\mathcal{O}(\Omega)$ such that $\Omega$ is H-convex.

Theorem 7.- _Let_ I _be a_ b-_ideal of_ H; _if there exists a bounded family_ $(h_\alpha)$ _in_ I _such that_

$$\sup_\alpha |h_\alpha| \geqslant 1$$

_in_ $\Omega$ , _we have_ I = H.

_Proof_. We can easily find a bounded family $({}^s h)_{s\in\Omega}$ in I such that ${}^s h(s) \geqslant \frac{1}{2}$ . Multiplying ${}^s h$ by $1/{}^s h(s)$ , we obtain a bounded family; we therefore may assume that ${}^s h(s) = 1$. Then ${}^s h - 1$ vanishes at s and is bounded in H when s ranges over $\Omega$ . Using the decomposition property, one can write ${}^s h - 1 = \langle z-s, {}^s u\rangle$ , that is

$$(5.6.1) \qquad \langle z-s, {}^s u\rangle + {}^s h = 1,$$

where ${}^s u_1, \dots, {}^s u_n$ are bounded in H independently of $s \in \Omega$ . As $\Omega$ is spectral for z in H, we can find coefficients ${}^s u_1, \dots, {}^s u_n$ bounded in H independently of $s \in \complement\Omega$ such that

$$(5.6.2) \qquad \langle z-s, {}^s u\rangle = 1.$$

From (5.6.1) and (5.6.2), it is immediately seen that 0 is spectral for z in H modulo I. By virtue of Proposition 12 of Chapter III, this implies I = H.

Corollary 1 (L. Hörmander).- _Let_ $\delta$ _be a weight function on_ $\mathbf{C}^n$ _such that, up to equivalence_, $-\log\delta$ _is plurisubharmonic on_ $\Omega = \{\delta > 0\}$ _and let_ $f_1, \dots, f_m$ _belong to_ $\mathcal{O}(\delta)$. _The_ b-_ideal generated by_ $f_1, \dots, f_m$ _in_ $\mathcal{O}(\delta)$ _is equal to_ $\mathcal{O}(\delta)$ _if and only if there exist some positive integer_ N _and some_ $\varepsilon > 0$ _such that_

$$|f_1| + \dots + |f_m| \geqslant \varepsilon\,\delta^N .$$

_Proof_. If $\mathrm{idl}(f_1, \dots, f_m; \mathcal{O}(\delta)) = \mathcal{O}(\delta)$, we can find $g_1, \dots, g_m$ in $\mathcal{O}(\delta)$ such that

$$1 = f_1 g_1 + \dots + f_m g_m .$$

There exist a positive integer N and $\varepsilon > 0$ such that $\varepsilon\delta^N |g_j| \leqslant 1$ for $j = 1, \dots, m$. Therefore $|f_1| + \dots + |f_m| \geqslant \varepsilon\,\delta^N$. Conversely, assume that such a property holds. Using Theorem 5 of Chapter IV, we can find a bounded family $(g_\beta)$ in $\mathcal{O}(\delta)$ such that

$$\sup_\beta |g_\beta| \geqslant m/\varepsilon\delta^N.$$

Setting now $\alpha = (j, \beta)$ and $h_\alpha = f_j g_\beta$ , obviously $(h_\alpha)$ is a bounded family in $\mathrm{idl}(f_1, \dots, f_m; \mathcal{O}(\delta))$ and $\sup_\alpha |f_\alpha| \geqslant 1$. From Theorem 6 we then get $\mathrm{idl}(f_1, \dots, f_m; \mathcal{O}(\delta)) = \mathcal{O}(\delta)$.

Corollary 1 can be applied to algebras as $\mathcal{O}(\delta_\Omega)$ or $\mathcal{O}(\delta_{\Omega,\varphi})$ when $\Omega$ is a pseudoconvex open set in $\mathbf{C}^n$ and $\varphi$ a convex increasing mapping of $[0,\infty[$ into $[0,\infty[$ such that $\delta_{\Omega,\varphi}$ is a weight function.

Corollary 2.- Let $\Omega$ be a pseudoconvex open set in $\mathbf{C}^n$ and $\varphi$ be a non bounded convex increasing mapping of $[0,\infty[$ into $[0,\infty[$. Let also $f_1,\dots,f_m$ be elements of $\mathcal{O}(\Lambda_{\Omega,\varphi})$. Then $\operatorname{idl}(f_1,\dots,f_m;\mathcal{O}(\Lambda_{\Omega,\varphi})) = \mathcal{O}(\Lambda_{\Omega,\varphi})$ if and only if there exist positive numbers $C, c, \varepsilon$ such that

$$|f_1| + \dots + |f_m| \geqslant \varepsilon \exp(-C\varphi(-\log c\,\delta_\Omega)).$$

Proof. We have already seen in Section 1.5, that $\mathcal{O}(\Lambda_{\Omega,\varphi}) = \mathcal{O}((\tilde{\lambda}_c))$, where

$$\lambda_c = \exp(-\varphi(-\log c\,\delta_\Omega)).$$

As $\Omega$ is pseudoconvex, $-\log\delta_\Omega$ is plurisubharmonic in $\Omega$ and as $\varphi$ is convex, also $\varphi(-\log c\,\delta_\Omega) = -\log\lambda_c$ is. From Proposition 2 of Chapter IV, we know that $-\log\tilde{\lambda}_c$ is plurisubharmonic in $\Omega$. The condition of Corollary 2 is obviously necessary; conversely it implies that $|f_1| + \dots + |f_m| \geqslant \varepsilon\tilde{\lambda}_c^N$ for some positive integer N and some $\varepsilon>0$, $c>0$. Using Corollary 2, we see that $\operatorname{idl}(f_1,\dots,f_m;\mathcal{O}(\tilde{\lambda}_c))$ contains 1. Hence $\operatorname{idl}(f_1,\dots,f_m;\mathcal{O}(\Lambda_{\Omega,\varphi}))$ also contains 1 and the proof is complete.

A generalization of Theorem 7 is the following

Theorem 8.- Let I be a b-ideal of H and $g\in H$; if there exists a bounded family $(h_\alpha)$ in I such that

$$\sup_\alpha |h_\alpha| \geqslant |g|$$

in $\Omega$, then $g^k\in I$ for some positive integer k.

Proof. We can find a bounded family $({}^sh)_{s\in\Omega}$ in I such that ${}^sh(s) = g(s)$ for every $s\in\Omega$. Using the decomposition property, we write

$$g - {}^sh = \langle z-s, {}^su\rangle,$$

where ${}^su_1,\dots,{}^su_n$ are bounded in H independently of $s\in\Omega$.

We introduce the algebra $H_1 = H[X]$ of polynomials with coefficients in H, equipped with the natural boundedness described in Section 2.5, and the b-ideal

$$I_1 = I + (1-gX)H_1.$$

Clearly

$$\langle z-s, {}^suX\rangle + {}^shX + (1-gX) = 1,$$

and 0 is spectral for z in $H_1$ modulo $I_1$. Therefore $I_1 = H_1$ and $1-gX$ is invertible modulo $I[X]$. This implies that g is nilpotent modulo I and that $g^k\in I$ for some positive integer k.

Reasoning like for Corollaries 1 and 2, we deduce from Theorem 8

Corollary 3 (J.J. Kelleher, B.A. Taylor, I. Cnop).- Let $\delta$ be a weight function on $\mathbf{C}^n$ such that, up to equivalence, $-\log \delta$ is plurisubharmonic. Assume that $f_1, \ldots, f_m$, $g$ are functions of $\mathcal{O}(\delta)$ such that

$$|f_1| + \ldots + |f_m| \geqslant \varepsilon \, \delta^N \, |g| \, ,$$

for some positive integer $N$ and some $\varepsilon > 0$. Then there exists a positive integer $k$ such that $g^k$ belongs to the b-ideal of $\mathcal{O}(\delta)$ generated by $f_1, \ldots, f_m$.

Corollary 4.- Let $\Omega$ be a pseudoconvex open set in $\mathbf{C}^n$ and $\varphi$ be a non bounded convex increasing mapping of $[0, \infty[$ into $[0, \infty[$. Assume that $f_1, \ldots, f_m$, $g$ are functions of $\mathcal{O}(\Lambda_{\Omega,\varphi})$ such that

$$|f_1| + \ldots + |f_m| \geqslant \varepsilon \exp(-C \varphi(-\log c \, \delta_\Omega)) \, |g|$$

for some positive numbers $C$, $c$, $\varepsilon$. Then there exists a positive integer $k$ such that $g^k$ belongs to the b-ideal of $\mathcal{O}(\Lambda_{\Omega,\varphi})$ generated by $f_1, \ldots, f_m$.

## Notes

The decomposition property for algebras of holomorphic functions, equipped with boundednesses, has been introduced by the author (2) and used to study algebras $\mathcal{O}(\delta)$ in the one-dimensional case. Similar ideas are developed here in the n-dimensional case. Theorems 1 and 2 have been proved by the author (3), (4) by means of the double complex $L_r^t$ of Section 4.2. Diagram chasing was first used in this context by L. Hörmander (3) to prove Corollary 1 of Section 5.6 and by J.J. Kelleher and B.A. Taylor (1) to prove Corollary 3. The method adopted here, based on the holomorphic functional calculus modulo a b-ideal is due to L. Waelbroeck. Corollary 4 of Section 5.6 has also been obtained by J.J. Kelleher and B.A. Taylor in the case where $\Omega = \mathbf{C}^n$; our proof, based on plurisubharmonic regularization, is simpler.

# CHAPTER VI

## APPROXIMATION THEOREMS

We define the hull of a compact subset K of a given domain $\Omega$ with respect to a given algebra H of holomorphic functions on $\Omega$. In the case when $\Omega$ is $\mathbf{C}^n$ and H is the algebra of polynomials, we characterize the polynomially convex hull of K by means of spectral theory of Banach algebras and give a short proof of the so-called Oka-Weil Theorem. Using the results of Chapters IV and V, we discuss the case when H is the algebra of all holomorphic functions on a pseudoconvex domain $\Omega$ or, more generally, a subalgebra H with decomposition property such that $\Omega$ is H-convex. The results are applied to approximate holomorphic functions on the neighbourhood of a compact set, to study Runge domains, Runge pairs and generalize Runge property. We further extend the theory to approximation with growth. We consider a weight function $\delta$ and discuss density in $\mathcal{O}(\delta)$ of a subalgebra H with decomposition property; in particular, when $-\log\delta$ is plurisubharmonic in $\{\delta > 0\}$, density of H in $\mathcal{O}(\delta)$ is equivalent to the following convexity hypothesis: up to equivalence, $1/\delta$ is the supremum of moduli of functions of H. Another equivalent condition is given when H is equal to some $\mathcal{O}(\Delta')$, in particular when H is the algebra of polynomials, and examples of such a situation are considered, in connection with algebras $\mathcal{O}(e^{-|z|^\alpha})$, $\mathcal{O}(\delta_{\Omega,\varphi})$, $\mathcal{O}(\Lambda_\varphi)$ for instance.

### 6.1.- Approximation on compact sets

In this section, we consider an algebra H of holomorphic functions on a given open set $\Omega$. If K is a compact set in $\Omega$, we define the H-convex hull of K as the set $K_H$ of all points s in $\Omega$ such that

$$|f(s)| \leqslant \sup_{\zeta \in K} |f(\zeta)|$$

for every function f in H; we say that K is H-convex if $\widehat{K}_H = K$. When $\Omega = \mathbf{C}^n$ and H is the algebra of polynomials, we write $\widehat{K}_H = \widehat{K}_P$ ; it is called the polynomially convex hull of K; if $K = \widehat{K}_P$ we say that K is polynomially convex. When H is the algebra $\mathcal{O}(\Omega)$ of all holomorphic functions on $\Omega$ , we write $\widehat{K}_\Omega$ instead of $\widehat{K}_{\mathcal{O}(\Omega)}$.

Obviously $\widehat{K}_H$ is always closed in $\Omega$ . If a compact set K is H-convex, a fundamental system of neighbourhoods of K consists of subsets of $\Omega$ defined by inequalities

$$|f_1| < 1, \ldots, |f_m| < 1 ,$$

where $f_1, \ldots, f_m$ belong to H.

For suitable algebras H, we may characterize the H-convex hull by means of spectral theory. As a direct application of the theory of Banach algebras, we recall the following statement:

Proposition 1.- Let K be a compact set in $\mathbf{C}^n$; if $\overline{P}_K$ denotes the uniform closure on K of polynomials, we have $\widehat{K}_P = \mathrm{sp}(z; \overline{P}_K)$.

Proof. First $\widehat{K}_P$ is contained in $\mathrm{sp}(z; \overline{P}_K)$. If s belongs to $\widehat{K}_P$, consider the multiplicative linear form $\chi : p \mapsto p(s)$; it follows from the definition of $\widehat{K}_P$ that $\chi$ is continuous on the algebra of polynomials equipped with the uniform norm on K; thus $\chi$ can be continued to $\overline{P}_K$ and as

$$s = (\chi(z_1), \ldots, \chi(z_n)) ,$$

clearly s belongs to $\mathrm{sp}(z; \overline{P}_K)$.

We now show that $\widehat{K}_P$ contains $\mathrm{sp}(z; \overline{P}_K)$. If s belongs to $\mathrm{sp}(z; \overline{P}_K)$, there exists some multiplicative linear form $\chi$ on $\overline{H}_K$ such that $s = (\chi(z_1), \ldots, \chi(z_n))$. Then

$$p(s) = p(\chi(z_1), \ldots, \chi(z_n)) = \chi(p).$$

But the norm of $\chi$ is bounded by 1 and

$$|p(s)| \leqslant \sup_{\zeta \in K} |p(\zeta)| .$$

Therefore s belongs to $\widehat{K}_P$.

Corollary 1 (Oka-Weil).- Let K be a polynomially convex compact set in $\mathbf{C}^n$ and f an holomorphic function on a neighbourhood of K; then f is a uniform limit on K of polynomials.

Proof. As $K = \mathrm{sp}(z; \overline{P}_K)$, the holomorphic functional calculus enables us to define

$f[z]$ in the Banach algebra $\overline{P}_K$. From Proposition 11 of Chapter III, we get $f[z](s) = f(s)$ for every point $s$ in $K$. Therefore $f[z]$ is the restriction of $f$ to $K$ and the statement is proved.

We want to extend Proposition 1 to more general algebras $H$. We have used the polynomials only when writing $p(\chi(z)) = \chi(p)$. Suppose now that $H$ is the algebra $\mathcal{O}(\Omega)$ and let $\overline{\mathcal{O}(\Omega)}_K$ denote the uniform closure of $\mathcal{O}(\Omega)$ on a compact subset $K$ of $\Omega$. If $\chi$ is a multiplicative linear form on $\overline{\mathcal{O}(\Omega)}_K$, it is continuous on $\mathcal{O}(\Omega)$, when equipped with the topology of uniform convergence on $K$. Then $\chi$ is a bounded multiplicative linear form on $\mathcal{O}(\Omega)$. When $\Omega$ is pseudoconvex, from Proposition 6 of Chapter IV, we get $\chi(z) \in \Omega$ and $\chi(f) = f(\chi(z))$. Therefore

Proposition 2.- Let $K$ be a compact subset of a pseudoconvex open set $\Omega$; then $\hat{K}_\Omega = sp(z; \overline{\mathcal{O}(\Omega)}_K)$.

Further

Corollary 2.- Let $K$ be an $\mathcal{O}(\Omega)$-convex compact subset of a pseudoconvex open set $\Omega$ and $f$ an holomorphic function on a neighbourhood of $K$; then $f$ is a uniform limit on $K$ of polynomials.

More generally, if $\overline{H}_K$ denotes the uniform closure on $K$ of functions of $H$, we have

Proposition 3.- Let $\Omega$ be an open set in $\mathbf{C}^n$ and $H$ a subalgebra with decomposition property of $\mathcal{O}(\Omega)$ such that $\Omega$ is $H$-convex; for every compact subset $K$ of $\Omega$, we have $\hat{K}_H = sp(z; \overline{H}_K)$.

It is an easy consequence of Proposition 5 of Chapter V. From Proposition 3 we deduce

Corollary 3.- We keep the assumptions of Proposition 3. Let $K$ be a $H$-convex compact subset of $\Omega$ and $f$ an holomorphic function on a neighbourhood of $K$; then $f$ is a uniform limit on $K$ of polynomials.

## 6.2.- Runge domains and generalizations

We first establish the following

Proposition 4.- Let $\Omega$ be an open set in $\mathbf{C}^n$; the following properties are equivalent:

(i) $\hat{K}_P \cap \Omega$ is compact in $\Omega$, for every compact subset $K$ of $\Omega$.

(ii) $\Omega$ is pseudoconvex and the polynomials are dense in $\mathcal{O}(\Omega)$.

(iii) $\hat{K}_P$ is contained in $\Omega$, for every compact subset $K$ of $\Omega$.

Proof. Assume that (i) holds. It is clear that the hull $\widehat{K}_\Omega$ of every compact subset K of $\Omega$ with respect to plurisubharmonic functions in $\Omega$ is contained in $\widehat{K}_P \cap \Omega$ ; therefore $\Omega$ is pseudoconvex. In view of Proposition 1, the spectrum of z in $\overline{P}_K$ is $\widehat{K}_P$ and the holomorphic functional calculus of Banach algebras gives a bounded linear mapping $\mathcal{O}(\widehat{K}_P) \to \overline{P}_K$. As $\widehat{K}_P \cap \Omega$ is closed in $\widehat{K}_P$, extending functions by zero, we get a bounded linear mapping $\mathcal{O}(\widehat{K}_P \cap \Omega) \to \mathcal{O}(\widehat{K}_P)$. Composing then with the natural mapping $\mathcal{O}(\Omega) \to \mathcal{O}(\widehat{K}_P \cap \Omega)$, we obtain a bounded linear mapping $\mathcal{O}(\Omega) \to \overline{P}_K$, which coincides with the restriction to K because $f[z](s) = f(s)$ for every $s \in K$, by virtue of Proposition 11 of Chapter III. Thus every function of $\mathcal{O}(\Omega)$ is uniformly approximable on K by polynomials and condition (ii) is proved.

Further (ii) implies (iii). Let $\chi$ be a multiplicative linear form on the Banach algebra $\overline{P}_K$. The restriction of functions to K maps $\mathcal{O}(\Omega)$ into $\overline{P}_K$ and $\chi$ defines a bounded multiplicative linear form on $\mathcal{O}(\Omega)$. As $\Omega$ is pseudoconvex, Proposition 6 of Chapter IV shows that $\chi(z)$ belongs to $\Omega$. Thus $\Omega$ contains the spectrum of z in $\overline{P}_K$ which is also $\widehat{K}_P$.

As (iii) obviously implies (i), the proof of Proposition 4 is complete.

If $\Omega$ satisfies the equivalent conditions of Proposition 4, we say that $\Omega$ is a Runge open set. Another characterization of Runge open sets is given by

Proposition 5.- Let $\Omega$ an open set in $\mathbf{C}^n$; $\Omega$ is Runge if and only if it is H-convex, where H is the algebra of polynomials equipped with the structure induced by $\mathcal{O}(\Omega)$.

Proof. We note that H has the decomposition property over $\Omega$ in view of Theorem 3 of Chapter V. If $\Omega$ is H-convex, it is spectral for z in $\widehat{H} = \overline{H}$; for every compact set K in $\Omega$, we have a natural morphism $\overline{H} \to \overline{P}_K$ and the spectrum of z in $\overline{P}_K$ is contained in $\Omega$. Then condition (iii) is fulfilled. Conversely, condition (ii) of Proposition 4 implies condition (ii) of Theorem 5 of Chapter V; if $\Omega$ is Runge, it is therefore H-convex.

Reasoning similar and using Proposition 2 instead of Proposition 1, we easily obtain

Proposition 6.- Let $\Omega$, $\Omega'$ be open sets in $\mathbf{C}^n$ such that $\Omega'$ is pseudoconvex and contains $\Omega$; the following properties are equivalent

(i) $\widehat{K}_{\mathcal{O}(\Omega')} \cap \Omega$ is compact in $\Omega$, for every compact subset K of $\Omega$.

(ii) $\Omega$ is pseudoconvex and $\mathcal{O}(\Omega')$ is dense in $\mathcal{O}(\Omega)$.

(iii) $\widehat{K}_{\mathcal{O}(\Omega')}$ is contained in $\Omega$, for every compact subset K of $\Omega$.

When the equivalent conditions of Proposition 6 are fulfilled, we say that $(\Omega, \Omega')$ is a Runge pair. We also have

Proposition 7.- Let $\Omega$, $\Omega'$ be open sets in $\mathbf{C}^n$ such that $\Omega'$ is pseudoconvex and contains $\Omega$ ; then $(\Omega, \Omega')$ is a Runge pair if and only if $\Omega$ is H-convex, where H is $\mathcal{O}(\Omega')$ equipped with the structure induced by $\mathcal{O}(\Omega)$.

We finally give a general statement including both Propositions 4 and 5 which is a consequence of Proposition 3.

Proposition 8.- Let $\Omega$, $\Omega'$ be open sets in $\mathbf{C}^n$ such that $\Omega'$ contains $\Omega$ and H be a subalgebra with decomposition property of $\mathcal{O}(\Omega')$ such that $\Omega'$ is H-convex; the following properties are equivalent:

(i) $\hat{K}_H \cap \Omega$ is compact in $\Omega$, for every compact subset K of $\Omega$.

(ii) $\Omega$ is pseudoconvex and H is dense in $\mathcal{O}(\Omega)$.

(iii) $\hat{K}_H$ is contained in $\Omega$, for every compact subset K of $\Omega$.

(iv) $\Omega$ is $H_1$-convex, where $H_1$ is the algebra H equipped with the structure induced by $\mathcal{O}(\Omega)$.

## 6.3.- Basic approximation theorem

We study now approximation with respect to some weight function $\delta$ . The convexity hypothesis $1/\chi_K = \sup |f_\alpha|$ of Section 6.1 will be replaced by the condition $1/\delta = \sup |f_\alpha|$, up to equivalence on $\delta$ .

Theorem 1.- Let $\delta$ be a weight function on $\mathbf{C}^n$ and $\Omega$ an open set such that $\Omega \supset \omega = \{\delta > 0\}$. We consider a subalgebra H of $\mathcal{O}(\delta) \cap \mathcal{O}(\Omega)$ containing the polynomials and assume that H has the decomposition property over $\Omega$ , when equipped with the structure induced by $\mathcal{O}(\delta)$. Then the following conditions are equivalent:

(i) There exists a bounded family $(f_\alpha)$ in H such that $\delta \geqslant (1/\sup|f_\alpha|)$ over $\omega$.

(i') There exists a family $(f_\alpha)$ in H such that, up to equivalence

$$1/\delta = \sup |f_\alpha|$$

over $\Omega$ .

(ii) $\bar{H} = \mathcal{O}(\delta)$ and, up to equivalence, $-\log\delta$ is plurisubharmonic in $\Omega$ .

(iii) $\delta$ is spectral for z in $\bar{H}$.

Proof. As $\mathcal{O}(\delta)$ is complete, we recall that $\bar{H} = \hat{H}$; the structure of $\bar{H}$ is not necessarily induced by $\mathcal{O}(\delta)$, but the identity mapping is a morphism of $\bar{H}$ into $\mathcal{O}(\delta)$.

Using Propositions 3 and 4 of Chapter V, we first prove that conditions (i), (i') and (iii) are equivalent. If (i) holds, let B be an absolutely convex bounded set in H such that $1 \in B$ and $f_\alpha \in B$ for every $\alpha$. We have $\delta \geqslant \delta_B$ and, as H has the decomposition property over $\omega$, clearly $\delta$ is the restriction to $\omega$ of some spectral function for z in $\bar{H}$. But (i) implies condition (i) of Theorem 5 in Chapter V and $\omega$ is H-convex. Therefore $\omega$ is spectral for z in $\bar{H}$ and also $\delta$ is. Further, if (iii) holds, as H has the decomposition property over $\Omega$, there exists some absolutely convex bounded set B in H such that $1 \in B$ and $\delta \geqslant \delta_B$ on $\Omega$. Then

$$1/\delta \leqslant \sup_{f \in B} |f| .$$

There also exist a positive integer N and a positive number M such that each $f \in B$ satisfies $|f| \delta^N \leqslant M$. Hence

$$\sup_{f \in B} |f| \leqslant 1/M \delta^N ,$$

and condition (i') follows. We finally observe that (i') implies trivially (i).

We now show that conditions(ii) and (iii) are equivalent. Assume (ii); as $\bar{H} = \mathcal{O}(\delta)$, we only have to prove that $\delta$ is spectral for z in $\mathcal{O}(\delta)$; as, up to equivalence, $-\log \delta$ is plurisubharmonic in $\omega$, this is nothing but Corollary 1 of Chapter IV. Conversely assume (iii). Using the natural morphism $\bar{H} \to \mathcal{O}(\delta)$, we see that $\delta$ is spectral for z in $\mathcal{O}(\delta)$. Thus, up to equivalence, $-\log \delta$ is plurisubharmonic in $\omega$. Moreover the holomorphic functional calculus gives a bounded linear mapping $\mathcal{O}(\delta) \to \bar{H}$, which coincides with the identity mapping; therefore $\bar{H} = \mathcal{O}(\delta)$.

## 6.4.- Approximation with growth

We consider the case when the subalgebra H of $\mathcal{O}(\delta)$ is equal to some $\mathcal{O}(\Delta')$, where $\Delta'$ is a directed set of weight functions such that each $\delta' \in \Delta'$ is larger than some function equivalent to $\delta$.

Theorem 2.- Assume that the open set $\{\delta' > 0\}$ is pseudoconvex for each $\delta' \in \Delta'$. The following conditions imply that $\mathcal{O}(\Delta')$ is dense in $\mathcal{O}(\delta)$:

(i) Up to equivalence, $1/\delta$ is the supremum on $\{\delta > 0\}$ of a family $(|f_\alpha|)$, where each $f_\alpha$ belongs to $\mathcal{O}(\Delta')$.

(ii) Up to equivalence, $1/\delta$ is the supremum on $\{\delta > 0\}$ of a family $(\pi_\alpha)$, where each $\pi_\alpha$ is a log-plurisubharmonic function in some $\mathcal{C}(\delta_\alpha)$ with $\delta_\alpha \in \Delta'$.

Proof. By virtue of Theorem 3 of Chapter V, the algebra $\mathcal{O}(\Delta')$, equipped with the structure induced by $\mathcal{O}(\delta)$, has the decomposition property over $\omega = \{\delta > 0\}$. The first part of Theorem 2 is therefore obvious: condition (i) is nothing but a reformulation of condition (i') of Theorem 1. Moreover (i) easily implies (ii). We only have to prove the converse statement. Assuming that (ii) holds, we define a weight function $\gamma_\alpha$ by

$$\gamma_\alpha = \operatorname{Min}\left(\left(1/\pi_\alpha\right)^\sim , \delta_0\right),$$

where $1/\pi_\alpha$ has been extended by 0 on the complement of $\Omega_\alpha = \{\delta_\alpha > 0\}$. As $\Omega_\alpha$ is pseudoconvex and $-\log(1/\pi_\alpha)$ plurisubharmonic in $\Omega_\alpha$, also $-\log(1/\pi_\alpha)^\sim$ and $-\log \gamma_\alpha$ are. Using then Theorem 5 of Chapter IV, for each $\alpha$ we can find a family $(f_{\alpha,\beta})_\beta$ in $\mathcal{O}(\gamma_\alpha)$ such that

$$1/\gamma_\alpha \leqslant \sup_\beta |f_{\alpha,\beta}| \leqslant M/\gamma_\alpha^N ,$$

where N is a positive integer and M a positive constant, both independent of $\alpha$. We may assume that

$$1/\delta \leqslant \sup_\alpha \pi_\alpha \leqslant 1/\varepsilon\delta^P$$

for some positive integer P and some $\varepsilon > 0$, and that $\varepsilon$ is such that $\varepsilon\delta^P$ is a weight function bounded by $\delta_0$. As $\pi_\alpha \leqslant 1/\gamma_\alpha$, obviously

$$1/\delta \leqslant \sup_{\alpha,\beta} |f_{\alpha,\beta}| .$$

As $1/\pi_\alpha \geqslant \varepsilon\delta^P$, we also have $\gamma_\alpha \geqslant \varepsilon\delta^P$ and

$$|f_{\alpha,\beta}| \leqslant M/\varepsilon^N\delta^{NP} .$$

Therefore the proof will be complete if each $f_{\alpha,\beta}$ belongs to $\mathcal{O}(\Delta')$; as $\pi_\alpha$ is in $\mathcal{C}(\delta_\alpha)$, it is easily shown that $1/\pi_\alpha \geqslant \varepsilon\delta^P$, where $\varepsilon$, P depend on $\alpha$ and that $\gamma_\alpha$ is larger than some function equivalent to $\delta_\alpha$; then $\mathcal{O}(\gamma_\alpha)$ is contained in $\mathcal{O}(\delta_\alpha)$.

In the particular case when $\mathcal{O}(\Delta')$ is the algebra $\mathcal{O}(\delta_0)$ of polynomials, we obtain

Corollary 4.- The following conditions imply that the polynomials are dense in $\mathcal{O}(\delta)$:

(i) Up to equivalence, $1/\delta$ is the supremum of a family of moduli of polynomials.

(ii) Up to equivalence, $1/\delta$ is the supremum of a family of log-plurisubharmonic functions with polynomial growth on $\mathbf{C}^n$.

We immediately list a few examples.

1) Let $\alpha$ be a positive number and $\delta = e^{-|z|^\alpha}$. Maybe $\delta$ is not exactly a weight function, but it is homothetic to some one. Obviously

$$1/e^{-|z|^\alpha} = \sup_{p \geqslant 0} \sum_{q=0}^{p} \frac{|z|^{\alpha q}}{q!} ,$$

and each

$$\sum_{q=0}^{p} \frac{|z|^{\alpha q}}{q!}$$

is log-plurisubharmonic on $\mathbf{C}^n$ and has polynomial growth at infinity. Therefore the polynomials are dense in the algebra of entire functions of order $\alpha$.

2) Consider in $\mathbf{C}^n$ the polyedron $\omega$ defined by inequalities

$$|p_1| < 1, \ldots, |p_m| < 1 ,$$

where $p_1, \ldots, p_m$ are polynomials; if d denotes the distance to the complement of the unit disc in the complex plane, we set, when $\overline{\omega}$ is supposed to be compact

$$\delta = \mathrm{Min}\,(d \circ p_1, \ldots, d \circ p_m).$$

Obviously $\delta$ is homothetic to some weight function. It is easily seen that $\omega$ is exactly the set $\{\delta > 0\}$. Moreover

$$1/d = \sup_{|\zeta|>1} |(z-\zeta)^{-1}| ,$$

and each $(z-\zeta)^{-1}$, with $|\zeta| > 1$, is a uniform limit on the unit disc of polynomials. Then we can find a family $(q_\alpha)$ of polynomials such that

$$1/d = \sup_{\alpha} |q_\alpha|$$

on the unit disc. Therefore

$$1/\delta = \sup_{\alpha, i} |q_\alpha \circ p_i|$$

on $\omega$, and the polynomials are dense in $\mathcal{O}(\delta)$.

We can obtain through this method a new proof of the Oka-Weil Theorem. Assume that f is holomorphic on an open neighbourhood $\Omega$ of some polynomially convex compact subset K of $\mathbf{C}^n$. We can find a polynomial polyedron $\omega$ such that $K \subset \omega \subset \overline{\omega} \subset \Omega$. If $\delta$ is the function defined above, as f is bounded on $\overline{\omega}$, obviously f belongs to $\mathcal{O}(\delta)$. Then f is the limit in $\mathcal{O}(\delta)$ of polynomials; but convergence in $\mathcal{O}(\delta)$ implies uniform convergence on K.

Instead of a polynomial polyedron, we may consider a polyedron $\omega$ defined by inequalities

$$|f_1| < 1, \ldots, |f_m| < 1 ,$$

where $f_1, \ldots, f_m$ are holomorphic functions on a pseudoconvex open set $\Omega$ containing $\omega$. We thus obtain a new proof of Corollary 2.

3) Let $\Omega$ be a pseudoconvex open set in $\mathbf{C}^n$ and $\varphi$ a convex increasing mapping of $[0, +\infty[$ into $[0, +\infty[$ such that

$$\delta_{\Omega,\varphi} = \exp(-\varphi(-\log \delta_\Omega))$$

is equivalent to some weight function. For each positive integer p, let $\varphi_p$ be equal to $\varphi$ on $[0, p]$ and to some affine function tangent to $\varphi$ at p on $[p, +\infty[$ . It is easily seen that

$$\delta_{\Omega,\varphi_p}^{-1} = \exp(\varphi_p(-\log \delta_\Omega))$$

belongs to $\mathcal{C}(\delta_\Omega)$ and that $\log(\delta_{\Omega,\varphi_p}^{-1})$ is plurisubharmonic in $\Omega$ . As

$$\delta_{\Omega,\varphi}^{-1} = \sup_p \delta_{\Omega,\varphi_p}^{-1}$$

using Theorem (ii), we see that $\mathcal{O}(\delta_\Omega)$ is dense in $\mathcal{O}(\delta_{\Omega,\varphi})$. When $\varphi$ varies, we obtain that $\mathcal{O}(\delta_\Omega)$ is also dense in $\mathcal{O}(\Omega)$.

4) We again consider a pseudoconvex open set $\Omega$ in $\mathbf{C}^n$. Let $\varphi$ be a non bounded convex increasing mapping of $[0, +\infty[$ into $[0, +\infty[$ and $\mathcal{O}(\Lambda_{\Omega,\varphi})$ be the algebra defined in Section 1.5. We recall that $\Lambda_{\Omega,\varphi}$ is the set of all

$$\lambda_c = \exp(-\varphi(-\log c\,\delta_\Omega)).$$

Fix $c>0$; choose $\varphi_p$ as in example 3) and set

$$\pi_p = \exp(\varphi_p(-\log c\,\delta_\Omega)).$$

We see that $\pi_p$ belongs to $\mathcal{C}(\delta_\Omega)$ and that $\log \pi_p = \varphi_p(-\log \delta_\Omega - \log c)$ is plurisubharmonic in $\Omega$ . Moreover

$$1/\lambda_c = \sup_p \pi_p$$

and

$$\tilde{\lambda}_c = (1/\sup_p \pi_p)^{\sim} .$$

Then $\mathcal{O}(\delta_\Omega)$ is dense in $\mathcal{O}(\tilde{\lambda}_c)$. As the property is valid for every $c > 0$, the algebra $\mathcal{O}(\delta_\Omega)$ is dense in $\mathcal{O}(\Lambda_{\Omega,\varphi})$.

## Notes

The results exposed in Section 6.1, except Proposition 3 and Corollary 3, and in Section 6.2, except Proposition 8, are classical and most of them are due to K. Oka (1); applying here spectral theory, we obtain very short proofs. Approximation theorems with growth are developments of ideas of the author (3), (4).

# CHAPTER VII

# FILTRATIONS

We have studied in Chapters IV to VI the b-algebra $\mathcal{O}(\delta)$, when $\delta$ is a weight function. Here we are interested in the polynormed algebra $\mathcal{O}(\delta)$; instead of taking isomorphisms, we consider the structure given by the sequence $({}_N\mathcal{O}(\delta))$ of Banach spaces. In this context, we apply a precise spectral theorem to prove properties of plurisubharmonic functions on a pseudoconvex domain: for instance, if a plurisubharmonic $\varphi$ is such that $e^{-\varphi}$ is a weight function, it is the supremum of a family $1/n_\beta \log |g_\beta|$, where each $n_\beta$ is a positive integer and each $g_\beta$ is holomorphic. We also study approximation of functions of $\mathcal{O}(\delta)$ by functions of a subalgebra and refine theorems of Chapter VI. Finally a strict concept of polynomial convexity for open sets is introduced and compared with usual polynormed convexity.

## 7.1.- Filtrated b-algebras

Let $\underline{E} = (E, ({}_NE)_{N\in \mathbf{Z}})$ be a polynormed vector space defined by a covering indexed by $\mathbf{Z}$, such that each ${}_NE$ is a Banach space and the identity ${}_NE \to {}_{N+1}E$ has norm $\leqslant 1$; we say that $\underline{E}$ is a filtrated b-space. If B is a bounded set in $\underline{E}$, the smallest $N\in \mathbf{Z}$ such that B is bounded in ${}_NE$ is called the filtration of B and denoted by $\nu(B)$; if B is not bounded, we set $\nu(B) = +\infty$. The filtration of an element $x\in E$ is the filtration of $\{x\}$; it is denoted by $\nu(x)$.

Consider a linear mapping u of a filtrated b-space E into another F. We say that u has a finite filtration if there exists some $k\in \mathbf{Z}$ such that u is continuous from ${}_NE$ into ${}_{N+k}F$ for every $N\in \mathbf{Z}$ or, equivalently, if

$$\nu(u(B)) \leqslant \nu(B) + k ,$$

for every bounded set B in E. The smallest k such that the property holds is the filtration $\nu(u)$ of u.

Let F be a linear subspace of a filtrated b-space E. We naturally equip $\overline{F}$ with a structure of filtrated b-space, by considering for each $N \in \mathbf{Z}$, the closure ${}_N\overline{F}$ of $F \cap {}_N E$ in ${}_N E$. The identity mappings $F \to \overline{F}$ and $\overline{F} \to E$ have non positive filtrations. We say that F is dense with filtration in E if F is dense in E and if the identity mapping $E \to \overline{F}$ has a finite filtration. This means that every element of ${}_N E$ is the limit in ${}_{N+k}E$ of elements of F, where k is independent of N.

A filtrated b-space A, fitted out with a structure of algebra is said to be a filtrated b-algebra if

$$ {}_N A \cdot {}_P A \subset {}_{N+P}A , $$

for all N, P in $\mathbf{Z}$, and the multiplication ${}_N A \times {}_P A \to {}_{N+P}A$ has norm $\leqslant 1$.

For instance $\mathcal{C}(\delta)$, $\mathcal{O}(\delta)$, $\mathcal{C}_r(\delta)$, $\mathcal{C}\mathcal{C}_r(\delta)$ are filtrated b-algebras, when $\delta$ is a weight function such that $\delta \leqslant 1$.

## 7.2.- Spectral theorem with filtration

Let A be a commutative filtrated b-algebra with unit element and $a_1, \dots, a_n$ be elements of A. We say that a non negative function $\delta$ on $\mathbf{C}^n$ is spectral for $a_1, \dots, a_n$ with filtration if there exists some positive integer k such that for every positive integer N one can find bounded mappings $u_o, u_1, \dots, u_n$ of $\mathbf{C}^n$ into ${}_{N+k}A$ satisfying

$$ (a_1 - z_1)u_1 + \dots + (a_n - z_n)u_n + \delta^N u_o = 1 $$

and

$$ \limsup_{N \to +\infty} \left( \sup_{s \in \mathbf{C}^n} \| u_i(s) \|_{{}_{N+k}A} \right)^{1/N} \leqslant 1 . $$

Theorem 1.- Let $\delta$ be a weight function bounded by 1 on $\mathbf{C}^n$ and $\Omega$ the set where $\delta$ does not vanish; then $\delta$ is spectral for z with filtration in $\mathcal{O}(\delta)$ if and only if $-\log \delta$ is plurisubharmonic in $\Omega$.

Proof. Clearly $\delta$ is spectral for z with filtration in $\mathcal{O}(\delta)$ if there exists some positive integer k such that for every positive integer N and every $s \in \mathbf{C}^n$ one can find functions $u_o(s) : \zeta \mapsto u_o(s;\zeta), \dots, u_n(s) : \zeta \mapsto u_n(s;\zeta)$ of $\mathcal{O}(\delta)$ satisfying

$$ (\zeta_1 - s_1)\, u_1(s;\zeta) + \dots + (\zeta_n - s_n)\, u_n(s;\zeta) + \delta^N(s)\, u_o(s;\zeta) = 1 \tag{7.2.1} $$

and

$$ \delta^{N+k}(\zeta)\, u_i(s;\zeta) = O(C_N) , \quad i = 0, \dots, n , \tag{7.2.2} $$

with

$$\lim_{N\to+\infty} C_N^{1/N} = 1 .$$

First assume that such a property holds. Taking $s=\zeta$ in (7.2.1) we have

$$\delta^N(\zeta)\, u_o(\zeta;\zeta) = 1,$$

and using (7.2.2) we obtain

(7.2.3) $$1/\delta^N(\zeta) \leqslant \sup_{s\in\Omega} |u_o(s;\zeta)| \leqslant C_N/\delta^{N+k}(\zeta).$$

These inequalities can also be written

$$-\log\delta \leqslant \varphi_N \leqslant -(1+\tfrac{k}{N})\log\delta + \log C_N$$

with

$$\varphi_N = \sup_{s\in\Omega} (1/N \log|u_o(s)|).$$

Let then N tend to infinity; as $\log C_N$ and $\frac{k}{N}$ tend to zero, $\varphi_N$ uniformly converges to $-\log\delta$ on every compact subset of $\Omega$. As $\varphi_N$ is plurisubharmonic in $\Omega$, also $-\log\delta$ is.

We suppose now that $-\log\delta$ is plurisubharmonic in $\Omega$ and fix a positive integer N and $\varepsilon>0$. For a given s in $\Omega$, we consider the covering of $\mathbf{C}^n$ by

$${}^s\Omega_1 = \{\delta > (1+\varepsilon)\,\delta(s)\}$$

and

$${}^s\Omega_2 = \{\delta < (1+2\varepsilon)\,\delta(s)\} .$$

We choose a $\mathcal{E}^\infty$ partition of unit $({}^s\varphi_1, {}^s\varphi_2)$ subordinated to such a covering so that

(7.2.4) $$D\,{}^s\varphi_i = O(1/\varepsilon\delta(s))$$

for $i=1,2$ and every derivation D of order 1. For instance let $\rho$ be a $\mathcal{E}^\infty$ increasing mapping of $\mathbf{R}$ into $\mathbf{R}$ such that $\rho(x)=0$ for $x\leqslant\frac{2}{3}$ and $\rho(x)=1$ for $x\geqslant 1$. Using Proposition 4 of Chapter I, we find a $\mathcal{E}^\infty$ weight function $\delta'$ such that

$$\delta \leqslant \delta' \leqslant (1+\varepsilon/4)\,\delta$$

and that $D\delta'$ is uniformly bounded for every derivative D of order 1. Then

$${}^s\varphi_1 : \zeta \mapsto \rho\Big(2\,\frac{\delta'(\zeta) - (1+\varepsilon)\,\delta(s)}{\varepsilon\,\delta(s)}\Big)$$

and ${}^s\varphi_2 = 1 - {}^s\varphi_1$ satisfy the required properties, for $\varepsilon$ small enough.

As $-\log\delta$ is plurisubharmonic in $\Omega$, the open set $\omega = \{\delta>\delta(s)\}$ is pseudoconvex. In view of Theorem 1 of Chapter IV, we know that $\omega$ is spectral for z in $\mathcal{O}(\delta_\omega)$; as s does not belong to $\omega$, we can find holomorphic functions ${}^sU_1,\ldots, {}^sU_n$ on $\omega$ such that

$$1 = \langle z-s, {}^sU \rangle ,$$

and

$$\delta_\omega^{N_o} \, | {}^sU_i | \leqslant M_o, \quad i=1,\dots,n,$$

where $N_o$, $M_o$ are universal constants. Moreover when $\zeta \in \Omega_1$, $\zeta' \notin \omega$ we have

$$\delta(\zeta) - \delta(\zeta') \geqslant \varepsilon\delta(s) \geqslant \varepsilon\delta(\zeta')$$

and thereby if $\varepsilon \leqslant 1$

$$\varepsilon\delta(\zeta) \leqslant \varepsilon\delta(\zeta') + \varepsilon(\delta(\zeta) - \delta(\zeta')) \leqslant 2(\delta(\zeta) - \delta(\zeta')) \leqslant 2\,|\zeta - \zeta'| .$$

Hence the distance from $\zeta$ to the complement of $\omega$ is at least $\varepsilon/2\ \delta(\zeta)$. If $\varepsilon$ is sufficiently small so that $\varepsilon\delta \leqslant \delta_o$, we therefore obtain $\delta_\omega(\zeta) \geqslant \varepsilon/2\ \delta(\zeta)$ and

$$\delta^{N_o} \, | {}^sU_i | \leqslant M_o(\varepsilon/2)^{-N_o} .$$

Extending ${}^sU_i$ by zero on $\Omega \cap \complement\, {}^s\Omega_1$ and using $1 = {}^s\varphi_1 + {}^s\varphi_2$, we have

$$1 = \langle z-s, {}^sU\, {}^s\varphi_1 \rangle + {}^s\varphi_2 ,$$

that is

$$1 = \langle z-s, {}^sv \rangle + \delta^N(s)\, {}^sv_o ,$$

if

$${}^sv_i = {}^sU_i\, {}^s\varphi_1, \quad i=1,\dots,n$$

and

$${}^sv_o = \delta^{-N}(s)\, {}^s\varphi_2 .$$

We easily see that

$$\delta^{N_o}\, {}^sv_i = O(\varepsilon^{-N_o}) , \quad i=1,\dots,n \tag{7.2.5}$$

and, as $\delta \leqslant (1+2\varepsilon)\,\delta(s)$ on the support of ${}^sv_o$, that

$$\delta^N\, {}^sv_o = O((1+2\varepsilon)^N). \tag{7.2.6}$$

Coefficients ${}^sv_o, {}^sv_1, \dots, {}^sv_n$ are not holomorphic in $\Omega$, but differentiable; as $\delta \leqslant (1+2\varepsilon)\,\delta(s)$ on the support of $d''\,{}^sv_i = U_i\, d''\,{}^s\varphi_1$, using (7.2.4) we obtain

$$\delta^{N+N_o+2}\, d''\,{}^sv_i = O(C_{N,\varepsilon}\, \delta^{N+1}(s)) , \quad i=1,\dots,n, \tag{7.2.7}$$

with

$$C_{N,\varepsilon} = \varepsilon^{-N_o-1}(1+2\varepsilon)^N .$$

Using Proposition 2 of Chapter I, it is easy to transform uniform estimates (7.2.5) to (7.2.7) into $L^2$-estimates. First considering $i=1,\dots,n$, for a sufficiently large positive integer $k'$, only depending on $n$, we obtain

$$\left( \int |d''\,{}^sv_i|^2\, \delta^{2(N+k')} d\lambda \right)^{1/2} = O(C_{N,\varepsilon}\, \delta^{N+1}(s)).$$

By virtue of Hörmander's d"-Lemma, there exists a locally integrable function ${}^s w_i$ on $\Omega$ such that $d'' \, {}^s w_i = d'' \, {}^s v_i$ and

$$\int |{}^s w_i|^2 \, \delta_o^4 \, \delta^{2(N+k')} \, d\lambda \leqslant \int |d'' \, {}^s v_i|^2 \, \delta^{2(N+k')} \, d\lambda .$$

Therefore

$$\left(\int |{}^s w_i|^2 \, \delta^{2(N+k'+2)} \, d\lambda\right)^{1/2} = O(C_{N,\varepsilon} \, \delta^{N+1}(s)).$$

We now set ${}^s u_i = {}^s v_i - {}^s w_i$ for $i = 1, \ldots, n$ and

$${}^s u_o = \delta(s)^{-N} (1 - \langle z - s, {}^s u \rangle ).$$

Obviously ${}^s u_1, \ldots, {}^s u_n$ are holomorphic in $\Omega$ and also ${}^s u_o$ is. Moreover

$${}^s u_o = {}^s v_o + \delta(s)^{-N} \langle z - s, {}^s w \rangle .$$

As $\delta(s) \, \delta \, |z - s|$ is bounded, we have

$$\delta(s)^{-N} \left(\int |\langle z - s, {}^s w \rangle| \, \delta^{2(N+k'+3)} \, d\lambda\right)^{1/2} = O(C_{N,\varepsilon}) .$$

From (7.2.6) we deduce a similar estimate for ${}^s v_o$; then ${}^s u_o$ also satisfies such an estimate. Using again Proposition 2 of Chapter I, we can transform this last $L^2$-estimate into a uniform one. For a sufficiently large positive integer k, only depending on n, we obtain

$$\delta^{N+k} \, {}^s u_o = O(C_{N,\varepsilon}).$$

Reasoning similar for $i = 1, \ldots, n$, we also have

$$\delta^{N+k} \, {}^s u_i = O(C_{N,\varepsilon}).$$

Choosing now $\varepsilon = 1/N$, we have

$$\lim_{N \to +\infty} (C_{N,\varepsilon})^{1/N} = 1 ,$$

and the statement is proved when s belongs to $\Omega$. When $\delta(s) = 0$, we can consider the coefficients $u_1(s), \ldots, u_n(s)$ given by Theorem 1 of Chapter IV; they satisfy

$$(z_1 - s_1) \, u_1(s) + \ldots + (z_n - s_n) \, u_n(s) + \delta^N(s) = 1$$

and

$$\delta^k \, u_i = O(1) , \quad i=1, \ldots, n ,$$

for some positive integer k; then we also have

$$\delta^{N+k} \, u_i = O(1) , \quad i=1, \ldots, n ,$$

and the proof is complete.

7.3.- Application to plurisubharmonic functions

A consequence of the methods developed in Section 7.2 is

Theorem 2.- Let $\delta$ be a Lipschitz non negative function over $\mathbf{C}^n$ such that $|z|\delta$ is bounded and $\Omega$ the set where $\delta$ does not vanish; the following conditions are equivalent:

(i) $-\log\delta$ is plurisubharmonic in $\Omega$ .

(ii) for every positive integer N, there exists a family $(f_\alpha)$ of holomorphic functions in $\Omega$ so that

$$1/\delta^N \leqslant \sup_\alpha |f_\alpha| \leqslant C_N/\delta^{N+k} ,$$

where k is independent of N and

$$\lim_{N\to+\infty} C_N^{1/N} = 1$$

(iii) there exist a family $(n_\beta)$ of positive integers and a family $(g_\beta)$ of holomorphic functions in $\Omega$ so that

$$-\log\delta = \sup_\beta 1/n_\beta \log|g_\beta| .$$

Proof. Assume that (i) holds. For a sufficiently small positive number $\lambda$ , clearly $\lambda\delta$ is a weight function such that $\lambda\delta \leqslant 1$. As $-\log\lambda\delta$ is plurisubharmonic in $\Omega$ , Theorem 1 shows that $\lambda\delta$ is spectral for z with filtration in $\mathcal{O}(\lambda\delta)$. Therefore (7.2.3) can be written

$$1/\lambda^N\delta^N \leqslant \sup_{s\in\Omega} |u_o(s)| \leqslant C_N/\lambda^{N+k}\,\delta^{N+k}$$

where each $u_o(s)$ is holomorphic in $\Omega$ and $\lim_{N\to+\infty} C_N^{1/N} = 1$. As

$$1/\delta^N \leqslant \sup_{s\in\Omega} |\lambda^{-N} u_o(s)| \leqslant \lambda^{-k} C_N/\delta^{N+k}$$

and

$$\lim_{N\to+\infty} (\lambda^{-k} C_N)^{1/N} = 1 ,$$

condition (ii) is proved.

Supposing now (ii), let $(f_{N,\alpha})_\alpha$ be a family of holomorphic functions in $\Omega$ such that

$$1/\delta^N \leqslant \sup_\alpha |f_{N,\alpha}| \leqslant C_N/\delta^{N+k}.$$

We immediately have

$$-\frac{N}{N+k}\log\delta - \log C_N \leqslant \sup_\alpha \frac{1}{N+k}\log|1/C_N\, f_{N,\alpha}| \leqslant -\log\delta ,$$

and condition (iii) follows with $\beta = (N, \alpha)$ and $g_\beta = 1/C_N \, f_{N,\alpha}$, as

$$\lim_{N \to +\infty} \left(-\frac{N}{N+k} \log \delta - \log C_N\right) = -\log \delta .$$

Finally (iii) obviously implies (i).

Proposition 1.- Let $\Omega$ be a pseudoconvex open set in $\mathbf{C}^n$ and $\varphi$ be a continuous plurisubharmonic function in $\Omega$. For every compact set K in $\Omega$, there exist a family $(n_\beta)$ of positive integers and a family $(g_\beta)$ of holomorphic functions in $\Omega$ such that

$$\varphi = \sup_\beta \, 1/n_\beta \, \log |g_\beta|$$

over K.

Proof. Let $\delta$ denote the function $e^{-\varphi}$ extended by zero on the complement of $\Omega$ and for every positive number $\lambda$ set

$$\delta_\lambda = 1/\lambda \, \mathrm{Min}(\delta_o, (\lambda\delta)^\sim).$$

Clearly $\delta_\lambda$ is Lipschitz over $\mathbf{C}^n$ and $|z|\,\delta_\lambda$ is bounded. In view of Proposition 2 of Chapter IV, $-\log \delta_\lambda$ is plurisubharmonic in $\Omega$. We have

$$\delta_\lambda(s) = \mathrm{Min}\left(\frac{\delta_o(s)}{\lambda}, \inf_{s' \in \mathbf{C}^n} \left(e^{-\varphi(s')} + \frac{|s'-s|}{\lambda}\right)\right),$$

where $e^{-\varphi(s')}$ is replaced by 0 when $s' \notin \Omega$. For a fixed s in $\Omega$, clearly $\delta(s)$ is the increasing limit of $\delta_\lambda(s)$ when $\lambda$ tends to zero. We therefore can uniformly approximate $\delta$ by functions $\delta_\lambda$ on every compact subset K of $\Omega$. Applying then Theorem 2 to $\delta_\lambda$, we easily obtain Proposition 1.

Corollary 1.- Let $\Omega$ be a pseudoconvex open set in $\mathbf{C}^n$; for every compact subset K of $\Omega$ we have $\hat{K}_\Omega = \hat{K}_{\mathcal{O}(\Omega)}$.

We have already seen that $\hat{K}_\Omega$ is the hull of K with respect to continuous plurisubharmonic functions in $\Omega$. From Proposition 1, it is also the hull of K with respect to functions $1/p \log|g|$ when p is a positive integer and g a function of $\mathcal{O}(\Omega)$ and we easily have $\hat{K}_\Omega = \hat{K}_{\mathcal{O}(\Omega)}$.

Using similar methods, we also obtain

Proposition 2.- Let $\varphi$ be a continuous plurisubharmonic function on $\mathbf{C}^n$ such that $e^\varphi$ has polynomial growth at infinity. For every compact set K in $\mathbf{C}^n$, there exist a family $(n_\beta)$ of positive integers and a family $(q_\beta)$ of polynomials such that

$$\varphi = \sup_\beta \, 1/n_\beta \, \log |q_\beta|$$

over K.

We only have to note that $\delta_\lambda$ has polynomial growth when $\delta$ has.

Corollary 2.- *For every compact subset* K *of* $\mathbf{C}^n$, *the polynomially convex hull* $\widehat{K}_p$ *of* K *is equal to the hull* $\widehat{K}_p$ *of* K *with respect to plurisubharmonic functions* $\varphi$ *on* $\mathbf{C}^n$ *such that* $e^\varphi$ *has polynomial growth at infinity*.

7.4.- *Approximation theorems with filtration*

We are able to state a new approximation result:

Theorem 3.- *Let* $\delta$ *be a weight function bounded by* 1 *on* $\mathbf{C}^n$ *and* $\Delta'$ *a directed set of weight functions such that each* $\delta' \in \Delta'$ *is larger than some function equivalent to* $\delta$. *We assume that* $\{\delta' > 0\}$ *is pseudoconvex for each* $\delta' \in \Delta'$ ; *the following conditions are equivalent*, *where* $\Omega = \{\delta > 0\}$ :

(i) *for every positive integer* N, *there exists a family* $(f_\alpha)$ *of functions of* $\mathcal{O}(\Delta')$ *so that*

$$1/\delta^N \leqslant \sup_\alpha |f_\alpha| \leqslant C_N/\delta^{N+k}$$

*on* $\Omega$ , *where* k *is independent of* N *and*

$$\lim_{N\to+\infty} C_N^{1/N} = 1 \ ;$$

(ii) *there exist a family* $(n_\beta)$ *of positive integers and a family* $(g_\beta)$ *of functions of* $\mathcal{O}(\Delta')$ *so that*

$$-\log\delta = \sup_\beta 1/n_\beta \log|g_\beta|$$

*on* $\Omega$ ;

(iii) $1/\delta$ *is the supremum on* $\Omega$ *of a family* $(\pi_\alpha)$, *where each* $\pi_\alpha$ *is a* log-*plurisubharmonic function in some* $\mathcal{T}(\delta_\alpha)$ *with* $\delta_\alpha \in \Delta'$ ;

(iv) $-\log\delta$ *is plurisubharmonic in* $\Omega$ *and* $\mathcal{O}(\Delta')$ *is dense in* $\mathcal{O}(\delta)$ *with filtration* ;

(v) $\delta$ *is spectral for* z *in* $\overline{\mathcal{O}(\Delta')}$ *with filtration*.

*Proof*. We only have to follow the proof of Theorem 2 to show that (i) implies (ii). Setting $\pi_\alpha = |g_\alpha|^{1/n_\alpha}$ in (ii), we immediately obtain (iii). If $\mathcal{O}(\Delta')$ is dense in $\mathcal{O}(\delta)$ with filtration there exists some positive integer k' such that every element of ${}_N\mathcal{O}(\delta)$ belongs to the closure ${}_{N+k'}\overline{\mathcal{O}(\Delta')}$ of $\mathcal{O}(\Delta') \cap {}_{N+k'}\mathcal{O}(\delta)$ in ${}_{N+k}\mathcal{O}(\delta)$, and the norm in ${}_N\mathcal{O}(\delta)$ is larger than the norm in ${}_{N+k'}\overline{\mathcal{O}(\Delta')}$ as $\delta \leqslant 1$. It is therefore easily seen that (iv) implies (v).

Assume that (v) holds. For every positive integer N and every $s \in \mathbf{C}^n$, one can find functions $u_0(s) : \zeta \mapsto u_0(s;\zeta), \ldots, u_n(s) : \zeta \mapsto u_n(s;\zeta)$ in ${}_{N+k}\overline{\mathcal{O}(\Delta')}$ satisfying (7.2.1) and (7.2.2). We easily obtain (7.2.3) that is:

$$1/\delta^N \leq \sup_{s \in \Omega} |u_o(s)| \leq C_N/\delta^{N+k}.$$

For every $\varepsilon \in ]0, 1]$, choose some $_\varepsilon u_o(s)$ in $\mathcal{O}(\Delta')$ such that

$$\delta^{N+k} \quad |_\varepsilon u_o(s) - u_o(s)| \leq \varepsilon \quad .$$

Thus

$$1/\delta^N \leq \sup_{\substack{s \in \Omega \\ \varepsilon \in ]0, 1]}} |_\varepsilon u_o(s)| \leq (C_N + 1)/\delta^{N+k}$$

and condition (i) is proved, as

$$\lim_{N \to +\infty} (C_N + 1)^{1/N} = \lim_{N \to +\infty} C_N^{1/N} = 1 .$$

We only have to show that $\mathcal{O}(\Delta')$ is dense in $\mathcal{O}(\delta)$ with filtration when (iii) is assumed. It is a consequence of Proposition 2 of Chapter I and the following

Lemma 1.- <u>Let $\delta$ be a locally integrable non negative function on an open set $\Omega$ of $\mathbb{C}^n$. We consider a sequence $(\Omega_p, \varphi_p)$, where $\Omega_p$ is a pseudoconvex open set of $\mathbb{C}^n$ containing $\Omega$ and $\varphi_p$ a plurisubharmonic function in $\Omega_p$, such that $\Omega_{p+1} \subset \Omega_p$ and $\varphi_{p+1} \geqslant \varphi_p$ on $\Omega_{p+1}$. Assume that $-\log \delta = \sup_p \varphi_p$ on $\Omega$. Then the closure of</u>

$$\bigcup_p \mathcal{O}(\Omega_p) \cap L^2(\delta_o^4 e^{-\varphi_p} d\lambda)$$

<u>in $L^2(\delta_o^4 \delta d\lambda)$ contains $\mathcal{O}(\Omega) \cap L^2(\delta_\Omega^{-2} \delta d\lambda)$, when $e^{-\varphi_o} \in L^1_{loc}(\Omega)$.</u>

<u>Proof.</u> We first define a sequence $(\alpha_q)$ of $\mathcal{C}^\infty$ functions with compact support on $\Omega$ such that

1) $0 \leqslant \alpha_q \leqslant 1$, $\alpha_q(s) = 1$ when $\delta_\Omega(s) \geqslant 2^{-q+1}$, $\alpha_q(s) = 0$ when $\delta_\Omega(s) \leqslant 2^{-q-1}$;

2) $\delta_\Omega D\alpha_q$ is uniformly bounded independently of q for every derivative D of order 1.

For instance, if $\rho$ is a $\mathcal{C}^\infty$ function on $\mathbb{C}^n$ with support in the unit ball and such that $\int \rho(s) d\lambda(s) = 1$ and $\rho_q(s) = 2^{2nq}\rho(2^q s)$ we may take $\alpha_q = \chi_q * \rho_{q+1}$ where $\chi_q$ is the characteristic function of the set $\{\delta_\Omega \geqslant 2^{-q}\}$.

Let f be a function of $\mathcal{O}(\Omega) \cap L^2(\delta_\Omega^{-2} \delta d\lambda)$. As $d''(f\alpha_q) = f d''\alpha_q$, we have

$$\int_\Omega |d''(f\alpha_q)|^2 \delta d\lambda = O\left( \int_{\delta_\Omega \geqslant 2^{-q-1}} |f|^2 \delta_\Omega^{-2} \delta d\lambda \right)$$

and the right hand side tends to zero when q tends to infinity. As $\delta$ is the limit of the decreasing sequence $e^{-\varphi_p}$, we can associate to each q some $p = p(q)$ such that

$$\varepsilon_q = \int_\Omega |d''(f\alpha_q)|^2 e^{-\varphi_{p(q)}} d\lambda$$

tends to zero when q tends to infinity. Using Hörmander's lemma, we can find a locally integrable function $g_q$ on $\Omega_{p(q)}$ such that $d''g_q = d''(f\alpha_q)$ and

$$(7.4.1) \qquad \int_{\Omega_{p(q)}} |g_q|^2 \delta_o^4 e^{-\varphi_{p(q)}} d\lambda \leqslant \varepsilon_q .$$

As $d''(f\alpha_q - g_q) = 0$, clearly $f_q = f\alpha_q - g_q$ is holomorphic in $\Omega_{p(q)}$; moreover $f_q$ belongs to $L^2(\delta_o^4 e^{-\varphi_{p(q)}}d\lambda)$. From (7.4.1) we obtain

$$\int_\Omega |g_q|^2 \delta_o^4 \delta d\lambda \leqslant \varepsilon_q ,$$

so that $g_q$ tends to zero in $L^2(\delta_o^4 \delta d\lambda)$. It is easily seen that f is the limit of the sequence $(f\alpha_q)$ in $L^2(\delta_\Omega^{-2} \delta d\lambda)$ and thereby in $L^2(\delta_o^4 \delta d\lambda)$. Finally f is the limit in $L^2(\delta_o^4 \delta d\lambda)$ of the sequence $f_q = f\alpha_q - g_q$ and the proof of Lemma 1 is complete.

<u>End of the proof of Theorem</u> 3. Let $\Omega_\alpha = \{\delta_\alpha > 0\}$ and $\gamma_\alpha = \mathrm{Min}((1/\pi_\alpha)^\sim, \varepsilon\delta_o)$ when $\varepsilon > 0$ is sufficiently small so that $\delta \geqslant \varepsilon\delta_o$. Then

$$-\log \delta = \sup_\alpha -\log \gamma_\alpha$$

and for every non negative integer N'

$$-\log \delta^{N'} = \sup -N' \log \gamma_\alpha .$$

Proposition 2 of Chapter IV shows that $-\log \gamma_\alpha$ is plurisubharmonic in $\Omega_\alpha$. As it is continuous, we can replace the family $(\pi_\alpha)$ by a sequence and apply Lemma 1 with $\delta^{N'}$ instead of $\delta$. Then

$$\bigcup_\alpha \mathcal{O}(\Omega_\alpha) \cap L^2(\gamma_\alpha^{N'+4} d\lambda)$$

in $L^2(\delta^{N'+4} d\lambda)$ contains $\mathcal{O}(\Omega) \cap L^2(\delta^{N'-2} d\lambda)$.

We only have now to use Proposition 2 of Chapter I. Each $\mathcal{O}(\Omega_\alpha) \cap L^2(\gamma_\alpha^{N'+4} d\lambda)$ is contained in $\mathcal{O}(\gamma_\alpha)$ and thereby in $\mathcal{O}(\delta_\alpha)$ and $\mathcal{O}(\Delta')$. Further, choosing $N' = 2N+2n+3$, clearly $\mathcal{O}(\Omega) \cap L^2(\delta^{N'-2} d\lambda)$ contains ${}_N\mathcal{O}(\delta)$, whereas $\mathcal{O}(\Omega) \cap L^2(\delta^{N'+4} d\lambda)$ is contained in ${}_{N+3n+7}\mathcal{O}(\delta)$.

<u>Remark.</u> We only have considered algebras and subalgebras. However the methods developed here can be applied to other cases. For instance

Proposition 3.- <u>Let $\Omega$ be a pseudoconvex open set in $\mathbf{C}^n$ and $\varphi$ a plurisubharmonic function on $\Omega$. The vector space $E_{\Omega,\varphi}$ of all holomorphic functions</u> f <u>in $\Omega$ such that</u>

$$\int |f|^2 \, e^{-\varphi} \, \delta_\Omega^N \, d\lambda < +\infty$$

for some positive integer N, is dense in $\mathcal{O}(\Omega)$, when $e^{-\varphi}$ is locally integrable.

Proof. Let f be an holomorphic function in $\Omega$ and K a compact subset of $\Omega$. We want to approximate f by functions of $E_{\Omega,\varphi}$ uniformly on K. We may assume that there exists some $\varepsilon > 0$ such that $K = \{\delta_\Omega \geq 2\varepsilon\}$. Clearly

$$\frac{\delta_\Omega}{(\delta_\Omega - \varepsilon)^+} = \sup_p \left( \sum_{q=0}^{p} \varepsilon^q \, \delta_\Omega^{-q} \right) .$$

As each $\delta_\Omega^{-q}$ is log-plurisubharmonic in $\Omega$, each

$$\varphi_p = \varphi + 2 \log \left( \sum_{q=0}^{p} \varepsilon^q \, \delta_\Omega^{-q} \right)$$

is plurisubharmonic in $\Omega$. Now let $-\log \delta = \sup_p \varphi_p$; obviously

$$\delta = e^{-\varphi} \left[ \frac{(\delta_\Omega - \varepsilon)^+}{\delta_\Omega} \right]^2 ,$$

and the set where $\delta$ does not vanish is exactly the set $\omega = \{\delta_\Omega > \varepsilon\}$. Lemma 1 shows that the closure of

$$\bigcup_p \mathcal{O}(\Omega) \cap L^2(\delta_o^4 \, e^{-\varphi_p} \, d\lambda)$$

in $L^2(\delta_o^4 \, \delta \, d\lambda)$ contains $\mathcal{O}(\omega) \cap L^2(\delta_\omega^{-2} \, \delta \, d\lambda)$. But $\delta_\omega^{-2} \delta$ is bounded in $\omega$ and the last vector space contains f. The statement is therefore proved as convergence in $L^2(\delta_o^4 \, \delta \, d\lambda)$ implies compact convergence on K and each $\mathcal{O}(\Omega) \cap L^2(\delta_o^4 \, e^{-\varphi_p} \, d\lambda)$ is contained in $E_{\Omega,\varphi}$.

Corollary 1 (L. Hörmander).- Let $\varphi$ be a plurisubharmonic function on $\Omega$; the vector space $E_\varphi$ of all entire functions f such that

$$\int |f|^2 \, e^{-\varphi} \, \delta_o^N \, d\lambda < +\infty$$

for some positive integer N, is dense in $\mathcal{O}(\mathbf{C}^n)$, when $e^{-\varphi}$ is locally integrable.

In the particular case when $\mathcal{O}(\Delta')$ is the algebra of polynomials, we deduce from Theorem 3 the following statement:

Corollary 2.- Let $\delta$ be a weight function bounded by 1 and $\Omega$ denote the set where $\delta$ does not vanish; the following conditions are equivalent:

(i) <u>for every positive integer</u> N, <u>there exists a family</u> $(p_\alpha)$ <u>of polynomials so that</u>

$$1/\delta^N \leqslant \sup_\alpha |p_\alpha| \leqslant C_N/\delta^{N+k}$$

<u>on</u> $\Omega$ (resp. <u>on</u> $\mathbf{C}^n$) <u>where</u> k <u>is independent of</u> N <u>and</u>

$$\lim_{N\to+\infty} C_N^{1/N} = 1 ;$$

(ii) <u>there exist a family</u> $(n_\beta)$ <u>of positive integers and a family</u> $(q_\beta)$ <u>of polynomials so that</u>

$$-\log \delta = \sup_\beta 1/n_\beta \log|q_\beta|$$

<u>on</u> $\Omega$ (resp. <u>on</u> $\mathbf{C}^n$);

(iii) $1/\delta$ <u>is the supremum on</u> $\Omega$ (resp. <u>on</u> $\mathbf{C}^n$) <u>of a family of</u> log - <u>plurisubharmonic functions with polynomial growth on</u> $\mathbf{C}^n$ ;

(iv) $-\log \delta$ <u>is plurisubharmonic in</u> $\Omega$ <u>and the polynomials are dense with filtration in</u> $\mathcal{O}(\delta)$ ;

(v) $\delta$ <u>is spectral for</u> z <u>in</u> $\overline{\mathcal{O}(\delta_o)}$ <u>with filtration</u>.

7.5.- <u>Polynomially convex open sets</u>

In Section 6.1. we have defined polynomial convexity for compact subsets of $\mathbf{C}^n$. Now let $\delta$ be a weight function on $\mathbf{C}^n$ such that $\delta \leqslant 1$; we say that $\delta$ is <u>polynomially convex</u> if $\delta$ satisfies the equivalent properties (i) to (v) of Corollary 2. An open subset $\Omega$ of $\mathbf{C}^n$ is said to be <u>polynomially convex</u> if $\delta_\Omega$ is. Every polynomially convex $\Omega$ is pseudoconvex; moreover, using property (ii) of Corollary 2, we obtain

Proposition 4.- <u>Every polynomially convex open set is Runge</u>.

The converse statement is not true. For instance, an open subset $\Omega$ of $\mathbf{C}$ is Runge if and only if $\dot{\mathbf{C}}\setminus\Omega$ is connected. Such open sets can be found which are not polynomially convex (*).

The infimum of a family of polynomially convex weight functions on $\mathbf{C}^n$ is polynomially convex. Therefore the interior of the intersection of a family of polynomially convex open subsets of $\mathbf{C}^n$ is polynomially convex. When $\delta$ is a weight function on $\mathbf{C}^n$ such that $\delta \leqslant 1$, we denote by $\hat{\delta}_P$ the smallest polynomially convex weight function which is larger than $\delta$. Similarly when $\Omega$ is an open set in $\mathbf{C}^n$, we denote by $\hat{\Omega}_P$ the smallest polynomially convex open subset containing $\Omega$. It is easily seen that $\hat{\Omega}_P = \{(\delta_\Omega)\hat{}_P > 0\}$.

Proposition 5.- a) $1/\hat{\delta}_P$ is the supremum of all log-plurisubharmonic functions $\pi$ with polynomial growth such that $\pi\delta \leqslant 1$.

b) $-\log \hat{\delta}_P$ is the supremum of all $p \log|q|$ where $p$ is a positive integer and $q$ a polynomial with $\log|q| \leqslant -\log \delta$.

Proof. Obviously $\exp(-\sup p \log|q|) \geqslant (\sup \pi)^{-1}$ and property (iii) of Corollary 2 shows that $\hat{\delta}_P \geqslant \exp(-\sup p \log|q|)$. Further $(\sup \pi)^{-1} \geqslant \inf(1/\pi)^{\sim}$, and $\delta_1 = \inf(1/\pi)^{\sim}$ is polynomially convex. But $\pi\delta \leqslant 1$ implies $\delta \leqslant (1/\pi)^{\sim}$. Therefore $\delta_1 \geqslant \delta$ and $\delta_1 \geqslant \hat{\delta}_P$.

Corollary 1.- Let $\omega$ be an open set in $\mathbf{C}^n$; assume that there exists a log-plurisubharmonic function with polynomial growth $\pi$ such that $\pi < 1$ on $\omega$ and $\pi > 1$ on $\complement\overline{\omega}$. Then $\omega$ is polynomially convex.

Proof. We only have to prove that $(\delta_\omega)\hat{}_P(s) = 0$ for every $s \in \complement\overline{\omega}$. As $\pi^p\delta_\omega \leqslant 1$ for every positive integer $p$, we have $1/\pi^p(s) \geqslant (\delta_\omega)\hat{}_P(s)$. When $p$ tends to infinity, we obtain the required property.

Let for instance $p_1, \ldots, p_m$ be polynomials. The open set $\{|p_1| + \ldots + |p_m| < 1\}$ or the open polyedron $|p_1| < 1, \ldots, |p_m| < 1$, are polynomially convex. As every polynomially convex compact set has a fundamental system of neighbourhoods composed of polynomial polyedrons, we obtain

Corollary 2.- Every polynomially compact subset of $\mathbf{C}^n$ has a fundamental system of neighbourhoods composed of polynomially convex open sets.

The converse statement is true; if $K$ is the intersection of some polynomially convex open sets, we easily see that $\widehat{K}_P \subset K$. But Corollary 2 of Section 7.3 shows that $\widehat{K}_P = \widehat{K}_P$ and $K$ is therefore polynomially convex.

In particular, the interior $\mathring{K}$ of every polynomially convex compact set is polynomially convex. There exist, however, polynomially convex open subsets $\Omega$ of $\mathbf{C}$ which are not bounded or such that $\overline{\Omega}$ is compact but not polynomially convex (*).

Notes

(*) For instance the open subset of $\mathbf{C}$ defined by $y < \exp(-|x|^\alpha)$ is polynomially convex if and only if $\alpha \geqslant 1$; writing $z = \rho e^{i\theta}$ with $\rho \geqslant 0$, $0 \leqslant |\theta| \leqslant \pi$, the open subset defined by $1 < \rho < 1 + \exp(-1/|\theta|^\alpha)$ is polynomially convex if and only if $\alpha \geqslant 1$; the complement in the unit disc of $[0, 1[$ is not polynomially convex. Such properties can be verified by using methods of the author (2).

Filtrations have been introduced by L. Waelbroeck [1] in his spectral theory of b-algebras; here the word is used with a more restrictive meaning. Proposition 1 has been stated by H.J. Bremermann [2] (see also P. Lelong [1]), but his proof is not correct. Theorem 2 is an improvement of the same result, as the properties are valid in the whole open set $\Omega$ ; equivalence between condition (i) and a condition similar to (iii) has also been obtained by N. Sibony by means of methods of Hartogs: the supremum of $1/n_\beta \log |g_\beta|$ is only replaced by its lowest upper semi-continuous majorant. The technique of Lemma 1 is due to B.A. Taylor; it has been used in this context by N. Sibony. The concept of a polynomially convex open set is a refinement of convexity with respect to $\mathcal{O}(\delta_o)$ introduced in Chapter V; polynomially convex open sets are $\mathcal{O}(\delta_o)$-convex and thereby Runge. The author does not know whether $\mathcal{O}(\delta_o)$-convex open sets which are not polynomially convex exist.

# BIBLIOGRAPHY


Arens, R. and Calderon, A.P. (1) Analytic functions of several Banach algebra elements, Ann. of Math. 62 (1955), 204-216.

Bochner, S. and Martin, W.T. (1) Functions of several complex variables.- Princeton University Press, 1948.

Bourbaki, N. (1) Théories spectrales, chapitre I, Algèbres normées.- Paris, Hermann, 1967.

Bremermann, H.J. (1) Uber die Aquivalenz der pseudokonvexen Gebiete und der Holomorphiegebiete im Raum von n komplexen Verändlichen, Math. Ann. 126 (1954), 63-91.

(2) Complex convexity, Trans. Amer. Math. Soc. 82 (1956), 17-51.

(3) Die charakteriesierung Rungescher Gebiete durch plurisubharmonic Funktionen, Math. Ann. 136 (1958), 173-186.

Buchwalter, H. (1) Topologies, bornologies et compactologies, thèse Univ. de Lyon, 1968.

Cartan, H. and Thullen, P. (1) Regularitäts-und Konvergenzbereiche, Math. Ann. 106 (1932), 175-177.

Cnop, I. (1) Un problème de spectre dans certaines algèbres de fonctions holomorphes à croissance tempérée, C. R. Acad. Sci. Paris A 270 (1970), 1690-1691.

(2) A theorem concerning holomorphic functions with bounded growth, thesis, Vrije Univ. Brussel, 1971.

(3) Spectral study of holomorphic functions with bounded growth, in publication, Ann. Inst. Fourier.

Cnop, I. and Ferrier, J.-P. (1) Existence de fonctions spectrales et densité pour les algèbres de fonctions holomorphes avec croissance, C. R. Acad. Sci. Paris A 273 (1971), 353-355.

Fantappié, L. (1) Ann. Mat. Pura Appl. 22 (1943).

Ferrier, J.-P. (1) Ensembles spectraux et approximation polynomiale pondérée, Bull. Soc. Math. France 96 (1968), 289-335.

(2) Séminaire sur les algèbres complètes, Lecture Notes in Mathematics 164.- Berlin, Springer Verlag, 1970.

(3) Approximation des fonctions holomorphes de plusieurs variables avec croissance, C. R. Acad. Sci. Paris A 271 (1970), 722-724.

(4) Approximation des fonctions holomorphes de plusieurs variables avec croissance, Ann. Inst. Fourier 22 (1972), 67-87.

(5) Sur la convexité holomorphe et les limites inductives d'algèbres

$\mathcal{O}(\delta)$, C. R. Acad. Sci. Paris A 272 (1971), 237-239.

Fuks, B.A. (1) Special chapters in the theory of analytic functions of several complex variables, Amer. Math. Soc., Translations of math. monographs 14 (1965).

Gelfand, I. (1) Normierte Ringe, Mat. Sb. 51 (1941).

Gunning, R.C. and Rossi, H. (1) Analytic functions of several complex variables, Prentice Hall series in modern analysis, 1965.

Hogbe-Nlend, H. (1) Théorie des bornologies et applications, Lecture Notes in Mathematics 213.- Berlin, Springer Verlag, 1971.

Hörmander, L. (1) $L^2$-estimates and existence theorems for the $\bar{\partial}$-operator, Acta Math. 113 (1965), 89-152.

(2) An introduction to complex analysis in several variables.- New-York, D. Van Nostand Company, 1966.

(3) Generators for some rings of analytic functions, Bull. Amer. Math. Soc. 73 (1967), 943-949.

Houzel, C. (editor) (1) Séminaire Banach(mimeographed), Ecole Normale Supérieure, Paris, 1963.

Kelleher, J.J. and Taylor, B.A. (1) Finitely generated ideals in rings of analytic functions, Math. Ann. 193 (1971), 225-237.

Kiselman, C.O. (1) On entire functions of exponential type and indicators of analytic functionals, Acta Math. 117 (1967), 1-35.

Lelong, P. (1) Fonctionnelles analytiques et fonctions entières (n variables), Séminaire de Mathématiques supérieures 28.- Montréal, Presses de l'Université, 1968.

Leray, J. (1) Fonction de variables complexes: sa représentation comme somme de puissances négatives de fonctions linéaires, R. C. Acad. Lincei, (8) 20 (1956), 589-590.

Malgrange, B. (1) Sur les systèmes différentiels à coefficients constants, Coll. Int. du C.N.R.S. 117 (1963), 113-122.

Martineau, A. (1) Indicatrices de croissance des fonctions entières de n variables, Invent. Math. 2 (1966), 81-86.

Narashiman, R. (1) Cohomology with bounds on complex spaces, Several complex variables, Lecture Notes in Mathematics 155, 141-150.- Springer Verlag, 1970.

Norguet, F. (1) Sur les domaines d'holomorphie des fonctions uniformes de plusieurs variables complexes, Bull. Soc. Math. France 82 (1954), 137-159.

Oka, K. (1) Sur les fonctions de plusieurs variables I, J. Sc. Hiroshima Univ. 6 (1936), 245-255.

(2) Sur les fonctions de plusieurs variables IX, Jap. J. Math. 23 (1953), 97-155.

Poly, J.B. (1) d''-cohomologie à croissance, Sém. P. Lelong 1967-68, Lecture Notes in Mathematics 103, 72-80.- Berlin, Springer Verlag, 1968.

Rubel, L. and Taylor, B.A. (1) A Fourier series method for entire and meromorphic functions, Bull. Soc. Math. France 96 (1968), 53-96.

Shilov, G.E. (1) On decomposition of a commutative normed ring in a direct sum of ideals, Amer. Math. Soc. Transl. 1 (1955), 37-48.

Sibony, N. (1) Approximation pondérée des fonctions holomorphes dans un ouvert de $\mathbf{C}^n$, Séminaire Choquet 1970/71, n° 21, 14 p.

Skoda, H. (1) Système fini ou infini de générateurs dans un espace de fonctions holomorphes avec poids, C. R. Acad. Sci. Paris A 273 (1971), 389-392.

Taylor, B.A. (1) On weighted polynomial approximation of entire functions, Pacific J. Math. 36 (1971), 523-539.

Waelbroeck, L. (1) Etude spectrale des algèbres complètes, Acad. Roy. Belg. Cl. Sci. Mém., 1960.

(2) Lectures in spectral theory, Dep. of Math., Yale Univ., 1963.

(3) About a spectral theorem, Function algebras (edit. by F. Birtel), 310-321.- Scott, Foresman and Co, 1965.

(4) Some theorems about bounded structures, J. Functional Analysis 1 (1967), 392-408.

(5) Topological vector spaces and algebras, Lecture Notes in Mathematics 230.- Berlin, Springer Verlag, 1971.

Weil, A. (1) L'intégrale de Cauchy et les fonctions de plusieurs variables, Math. Ann. 111 (1935).

Whitney, H. (1) On analytic extensions of differentiable functions defined on closed sets, Trans. Amer. Math. Soc. 36 (1934), 63-89.

(2) On ideals of differentiable functions, Amer. J. Math. 70 (1948), 635-658.